THE DIAMOND MAGNATES

BRIAN ROBERTS

The Diamond Magnates

ILLUSTRATED

HAMISH HAMILTON

LONDON

First published in Great Britain 1972
by Hamish Hamilton Ltd
90 *Great Russell Street London W.C.*1

SBN 241 02177 4

Printed in Great Britain by
Western Printing Services Ltd, Bristol

Contents

CONTENTS

ILLUSTRATIONS

Between pages 86 and 87

Between pages 182 and 183

For
Alexander Kiddie

PROLOGUE

A RATHER charming story is told about the way in which the first important diamond was found in South Africa.

In March 1867, John O'Reilly, a travelling trader dealing in animal skins, ostrich feathers and elephant tusks, called at a lonely farmstead near Hopetown in the Cape Colony. He was on his way south to settle his business affairs before setting out on a trading expedition. He had stopped at the farm, 'De Kalk', to rest his animals. Living on the farm was Schalk van Niekerk, a somewhat luckless farmer who was well known in the district as a collector of unusual, but apparently worthless, coloured stones. Van Niekerk's hobby was by no means uncommon. The countryside was bleak, amusements were few, and the pebbles, gathered mostly from the nearby Orange River, were sought after by local children who used them in their games of 'five stones'. However, van Niekerk took his hobby more seriously. He had been given a textbook on gem stones by a Government surveyor who had suggested that the area surrounding the farm was diamondiferous in nature; it was this, rather than the children's pastime, that had inspired his search for oddly shaped pebbles. Few of his neighbours seem to have shared his hopes.

When O'Reilly called at 'De Kalk' he was shown the farmer's collection. One stone in particular attracted his attention. Not only was it an unusual shape, but it was decidedly heavier than the others. Van Niekerk told him how it had been discovered.

In a small cottage on the farm lived a Boer family named Jacobs. Some time earlier, the fifteen-year-old son of this family, Erasmus Jacobs, had brought home the stone which he and his sister had found near the river. There is some doubt as to which of the children had actually picked up the stone, but they were playing with it on a day when van Niekerk called to see their mother. Spotting it, van Niekerk immediately became interested. It seemed to him to be different from any stone he had previously handled: the more he polished it, the more it glittered. Primed by his text-book, he took it to the window to see whether it would scratch glass. It did. He felt sure that, at last, he had discovered a diamond. When he told Mrs Jacobs this and offered to buy the stone, she laughed at the idea. It was just a stone picked up in the veld and she would not hear of being paid for it; he was welcome to it, she told him.

Although van Niekerk appears to have had the stone for some time before showing it to the visiting O'Reilly, he seems to have done nothing about it. However, encouraged by the trader's interest, he now decided to put his theory to the test. O'Reilly was on his way to the nearest large settlements—Hopetown and Colesberg—and van Niekerk asked him to take the stone with him to find out what it really was. This O'Reilly agreed to do.

In Hopetown the stone aroused little interest. The storekeepers to whom O'Reilly showed it laughed outright at the suggestion that it might be valuable. The traders of Colesberg were no less cynical. Not until O'Reilly showed the stone to Lorenzo Boyes, the Acting Commissioner of Colesberg, did he receive any encouragement. After examining it, Boyes was inclined to agree that it was a diamond. All the same, he was not altogether sure and decided to consult the local chemist who was reputed to know something about gem stones. The chemist was scornful. The stone, he said, was nothing more than a topaz: so sure of this was he that he offered to buy Lorenzo Boyes a new hat if he were proved wrong. To settle the matter, it was agreed to send the stone to the nearest geologist: Dr William Guybon Atherstone of Grahamstown.

Dr Atherstone was, in many ways, a remarkable man. Besides being the first surgeon in South Africa to perform an operation under anaesthetics (he amputated a leg with the use of ether in 1847) he was a keen naturalist and one of the country's pioneer geologists. He had, in fact, let it be known that he was prepared to test any mineral specimen sent to him and had been instrumental in exposing at least one geological hoax. The stone which he received from Lorenzo Boyes—in an unsealed envelope—presented him with a new challenge. After subjecting it to various tests, he con-sulted everyone in Grahamstown competent to pass an opinion and then delivered his verdict. He reported to Lorenzo Boyes that the stone had ruined every jeweller's file in Grahamstown; it was indeed a diamond; it weighed $21\frac{1}{4}$ carats. Faced with this superior knowledge, the Colesberg chemist capitulated and Lorenzo Boyes received a new hat.

The story of this diamond—which became known as the 'Eureka' —has been told many times. There are a number of variations and it is almost impossible to vouch for the details. Some of those who took part in the transaction tended to change their accounts over the years and, with other versions, allowance has to be made for second and third hand embroideries. However, of the basic facts there is no doubt. A stone picked up by a member of the Jacobs family was given to Schalk van Niekerk who, in turn, entrusted it to John O'Reilly to take to Colesberg. Lorenzo Boyes sent it to Dr Atherstone who judged it to be a diamond.

Atherstone did not return the stone to Lorenzo Boyes. Realising the significance of the find—'where that came from there must be lots more' he wrote—he informed the Colonial Secretary, who wired back suggesting that the stone be sent to Cape Town. At Cape Town Atherstone's judgement was confirmed by the French Consul, M. Heritte, and a diamond polisher, Louis Hond. The stone was valued at £500. Later, the Governor of the Cape bought the stone and a model of it was exhibited in Paris at the *Exposition Universelle de 1867*.

[2]

The 'Eureka' was probably not the first diamond to be found in South Africa. There are various unconfirmed stories of earlier finds and the Bushmen are said to have used diamonds to weight their hunting sticks. Nevertheless, it was undoubtedly the discovery made at 'De Kalk' which first attracted world attention to southern Africa's diamond potential.

Surprisingly, there was no immediate rush to the Cape Colony. The interest aroused by the 'Eureka' was somewhat tempered by the uncertainty of prospectors' rights at the Cape. The Government had a clause in land titles which reserved mineral rights to the Crown and this tended to dampen initial enthusiasm. Groups of local men, including Lorenzo Boyes and the Colesberg chemist, began to scour the Hopetown district, but the few stones that were unearthed did little to offset the doubts about possible Government restrictions.

Nor was the countryside itself especially inviting; indeed, in the words of one early traveller, 'Her Majesty possesses not, in all her empire, another strip of land so unlovely.' Those two wide, brown, sluggish rivers—the Orange and the Vaal—converge in an area remarkable chiefly for its desolation. Only the banks of the rivers are fringed with trees; for the rest the plains stretch away to the horizon, grey, cindery, scrubby. A stunted, three-foot acacia tree is a landmark, visible for miles. When it rains, the normally dry water-courses become raging torrents; when it blows, the dust is choking; when, as happens for most days of the year, the sun shines, it is like an oven. It was no wonder that neither the British nor the Cape authorities were anxious to embark on a costly survey of this inhospitable region.

But in June 1867 excitement at the Cape increased when John O'Reilly produced another diamond valued at between £150 and £200. The Government offered to buy this diamond and announced that it would take no action against prospectors for the time being. With this announcement, the searching parties became larger and hopeful speculators ranged over a wide area. The few experts who visited the region, however, were extremely sceptical.

In 1868, Professor James R. Gregory, a mineralogist of London University, was sent to the Cape by a Hatton Garden diamond merchant. He made what was considered a thorough inspection of the Hopetown district—including the beds of the Orange and Vaal rivers—and, on his return to England, published his findings in the *Geological Magazine*. His report was damning. 'During the time I was in South Africa,' he wrote, 'I made a very lengthy examination of the districts where diamonds are said to have been found, but saw no indication that would suggest the finding of diamonds or diamond bearing deposits in any of these localities. The geological character of that part of the country renders it impossible, with the knowledge in our present possession of diamond bearing rocks, that any could have been discovered there.' Privately he expressed the opinion that the few diamonds that had so far come to light had either been carried there by ostriches or had been deliberately placed in a desolate area in the hopes of attracting immigrants and improving land values. These findings were heatedly dismissed in South Africa. Doubts were expressed about Gregory's integrity and Dr Atherstone, incensed at being considered a dupe or an impostor, wrote a spirited reply to the *Geological Magazine* report.

The Gregory dispute was still raging when the next important event occurred, in March 1869. Once again it revolved around Schalk van Niekerk. With the discovery and sale of the 'Eureka', van Niekerk's reputation as a collector of unusual stones had received a tremendous boost. His hobby was now treated with respect, not only by his neighbours, but throughout the territory. Therefore it was inevitable that when a Griqua shepherd, working near the Orange River, picked up a stone which looked like a diamond, he was advised to take it to van Niekerk. As in the case of the discovery of the first diamond, versions of how this much larger stone reached van Niekerk vary: but reach him it did. He had no difficulty in recognising the value of the stone. He made the shepherd a handsome offer: 500 sheep, 10 oxen and a horse. These were assets which a Griqua could appreciate and the stone was duly exchanged for the livestock. Van Niekerk now knew enough to dispense with the services of a go-between. He took the stone to Hopetown and sold it to a trading firm for £11,200. Within a short space of time, both van Niekerk and the Griqua shepherd had become rich men.

It took the traders a little longer to realise their fortune. A group of speculators, hearing of the transaction, claimed that the stone had been discovered on ground over which they held a concession. They applied to the Courts to stop the diamond from being exported to London. The judgement in the case which followed went against the speculators and attracted world-wide attention to the Cape Supreme Court. The interest was not so much for the legal niceties involved as for the size and quality of the stone: it was a splendid white diamond of 83½ carats. Known as 'The Star of South Africa', it was eventually sold, cut and polished, to the Earl of Dudley in London for £30,000.

[3]

With the discovery of 'The Star of South Africa' doubts about the mineral potential of South Africa evaporated. The scientists had been confounded, the fortune hunters moved in.

South Africans led the way. In the Cape, Natal and the Orange Free State, tradesmen and farmers, civil servants and shop assistants, loaded their ox-wagons or set off on foot to prospect for diamonds. More finds were made and the numbers grew. For a while attention was concentrated on the Orange River: especially the Hopetown district where the 'Eureka' and 'The Star of South Africa' had been found. Prospecting in this area, however, proved disappointing and did not last long. Diamond finds near the Vaal River—where speculators had long since been established—although not spectacular, were more rewarding.

Towards the end of 1869, big finds were made at Klipdrift on the north bank of the Vaal; shortly afterwards a party from Natal struck it rich at Pniel on the opposite side of the river. Prospectors had worked both these places earlier but once serious digging began they quickly became the centre of operations, and a sizeable community developed around them. Other camps had been established further along the river (notably at Hebron on the northern bank) and diggers could be found working the Vaal and its tributaries in an area which stretched for a hundred or more miles. It has been estimated that the diamond yield from these camps eventually averaged £4,000 for every three hundred claims.

Men began to arrive in South Africa from all over the world: the purposeful, the hopeful and the adventurous banded together and trekked across the sandy veld to scratch for luck in the mud and gravel of the river beds. 'Butchers, bakers, sailors, tailors, lawyers, blacksmiths, masons, doctors, carpenters, clerks, gamblers, sextons, labourers, loafers, men of every pursuit and profession . . .' it is said, 'fell in line in the straggling procession to the Diamond Fields. Army officers broke furloughs to join the motley troupe, schoolboys ran away from school, and women, even of good families, could not be held back from joining their husbands and brothers in the long and wearisome journey to the banks of the Vaal.'

The initial trickle of foreigners was to swell over the years. They came from the goldfields of America and Australia, from the back streets of Bermondsey and Hamburg, from the bogs of Ireland and the ghettos of Russia. Their motives were as mixed as the company they kept: some came in hopes, others out of despair; the sieve of the experienced digger was filled by the spade of the enthusiastic novice. They were thrown together by chance—the chance of circumstance and the chance of fortune.

Claims were dug in the small hills, or koppies, which lined the river banks. Working close to the river made the washing of gravel relatively easy. The most difficult task was the removal of the iron-stone boulders which blocked many of the claims. Among these boulders was the gravel in which diamonds were found. Once the boulders were out of the way, the gravel—or 'stuff' as it was called—was loaded onto a cart and taken to the river to be sieved and washed. Various sieving processes evolved but the most popular was the 'cradle' which held two or three sieves of graded mesh. 'While the stuff is being rocked in this cradle,' wrote a visitor, 'one of the diggers pours buckets full of water into it. The gravel thus being thoroughly cleansed by this double process of sifting and washing, the large stones in the top sieve are hastily glanced over, to see if perchance any *big* diamond be amongst them, and the other sieve or sieves are taken out, and the contents emptied on to the "sorting table".... At this table, the digger either sits or lies, according as it has legs or rests on the ground, and quickly sorts over the stuff with the aid of an iron or wooden scraper. ... Diamonds, especially those of good quality, show out brilliantly, and can very seldom be missed on a sorting table; the larger gems, indeed, are often found in the sieve, or even in the act of digging.' Few finds went unheralded. A shout from a claim or a sorting table invariably brought the other diggers running.

For all their cosmopolitanism, the river diggings were surprisingly peaceful. They developed from the solid core of industrious colonists —who first established the camps—and, for the most part, remained remarkably respectable. The thuggery and rioting which was commonplace in the early days of the Californian and Australian goldfields played little part in the lives of the river diggers. Community spirit was strong and new arrivals were apt to be regarded with suspicion until they had proved their worth.

The work was hard, supplies uncertain, living conditions primitive and the climate veered to extremes. For a long time medical aid was practically non-existent and diggers brought down by 'camp-fever' (from which many died) had to rely on friends to nurse them. But, by and large, it was an invigorating life, free from the constraints of civilisation, and brightened by the magnificent setting and the prospects of uncovering a fortune. The tree-lined river, the massed white tents, the cheery *camaraderie* and optimism pervading the camps, produced an atmosphere which, for some, bordered on the idyllic. 'The quietude of these solitary regions,' wrote a digger in 1870, 'is now broken by the song of the happy digger as he goes forth in the morning to begin his day's labour among the huge boulders in his claim, where throughout the day, the sound of pick and shovel is heard, or perhaps the shout of a lucky digger who has just succeeded in finding a diamond.' Even those who found few diamonds were to look back on these days with nostalgia.

[4]

It was not the carefree diggers, but their squabbling neighbours who caused the first serious disturbance at the river diggings. In South Africa the most harmless seeming incident can produce a full scale political crisis: an event like the discovery of diamonds was therefore bound to have political repercussions. Not only was this inevitable but, considering the region in which diamonds were found, it appeared to be satanically ordained.

Few spots on the South African map were more vulnerable to political pressure than was the Vaal River region. Two Boer republics—the Orange Free State and the Transvaal—merged into the eastern and north-eastern fringes of the territory: to the south-west lay the British controlled Cape Colony. The confused pattern of territorial antagonisms was further aggravated by the fact that a large part of the area was claimed by a local Griqua tribe and that tribe was, itself, divided into opposing factions.

Dispute over the fuzzily mapped, unproductive region had been carried on in a desultory fashion for a number of years. However, with the discovery of diamonds, interest sharpened all round. South Africa was in financial doldrums; the possibility of underground riches whetted the appetites of governments as well as individuals. The Orange Free State quickly beaconed off a section of the region; the President of the Transvaal, M. W. Pretorius, laid claim to the northern bank of the Vaal; and the Griquas—advised by a some-what sinister Cape-born speculator—were urged to seek British arbitration. A protracted, racially factious wrangle resulted. It was finally settled (as such conflicts tended to be settled) by Britain annexing the disputed land.

Very early in the quarrel, a crisis arose which seriously threatened the operations of the river diggers. In June 1870, President Pretorius of the Transvaal, in an attempt to assert his authority, granted a diamond monopoly on the northern banks of the Vaal to three men for twenty-one years. The diggers regarded this as scandalous. They refused to acknowledge the concession and set up their own independent republic. At Klipdrift, an ex-able-bodied seaman, Stafford Parker, was elected President of the haphazardly constituted state and the diggers prepared to resist all outside interference.

President Parker, a picaresque figure with magnificent whiskers and a taste for elegant clothes, ruled his republic with a firmness which doubtless surprised some of his more frivolous subjects. A butcher was placed in charge of discipline and a bizarre set of punishments evolved: diamond thieves were flogged, prostitutes and drunks were put in stocks, card cheats were ducked in the river and pilferers placarded and paraded through the camp. Persistent offenders were expelled from the diggings and for extreme cases there was the 'spread-eagle': the criminal being pegged out in the dust and left to the mercy of the sun and flies. Another much dreaded punishment was that whereby hoodlums were dragged mercilessly through the river. Taxes were often collected at gun point.

Such measures were drastic but effective. President Parker's regime was to be remembered not so much for its defiance of authority as for its contribution to law and order. 'In justice to Mr Parker and his counsellors . . .', wrote a later visitor, 'I shall declare that one whisper of cruelty, other than these eccentric punishments, never reached my ears. They did many foolish acts and perhaps committed some wrongs. It may not be well to ask closely which way their revenue all went. But their procedure answered the demands upon it. No criminal lost his life and no honest man felt terror.'

The Diggers Republic lasted only a matter of months. A new British High Commissioner was appointed to the Cape Colony and a magistrate was sent to protect the interests of British subjects at the diamond diggings. Stafford Parker, a patriot at heart, capitulated. He told his followers that they could not fight Queen Victoria and refused to continue as President.

The collapse of the Diggers Republic coincided with a new turn of events. Towards the end of 1870 it became known that diamonds had been found on the farm of Dorstfontein and Bulfontein, some twenty-five miles south of Pniel. In January 1871 there was a rush away from the river to these new 'dry diggings'. Dorstfontein— which was to become better known as Dutoitspan—attracted the largest crowd, but there was no real centre for the feverish activities which followed this rush. Before long the diggers had invaded the adjacent farm, Vooruitzicht, owned by two brothers named De Beer.

Then, in July 1871, a party from Colesberg made what was to prove the most significant of all finds. On a small hill—about a mile from the De Beer farmhouse—a volcanic fissure, or 'pipe', was discovered in which were concentrated great quantities of diamonds.

The hillock became known as the Colesberg Koppie and the ensuing stampede there was called the New Rush: both these names, and the name of the owners of Vooruitzicht (De Beer) were to feature prominently in the legend of the diamond industry.

The 'dry diggings' were somewhat easier to work than those near the river. There were no large boulders to be removed and the top soil of soft rotten limestone was simply shifted with pick and shovel. A chronic shortage of water made it impossible to wash the soil which had to be broken up, or left to disintegrate in the sun, before being sieved. Otherwise the method of recovering diamonds was similar to that used at the river diggings. However, the finds were more rewarding. The stones unearthed at the dry diggings were to provide the foundations of South Africa's enormous diamond industry.

The four main camps—Dutoitspan, Bulfontein, De Beers and New Rush—quickly developed into tent towns. Canvas shops, offices, canteens and hotels sprang up overnight. There was a keen sense of rivalry between the camps and the merits of the four mines were hotly disputed. But New Rush gradually established its superiority. It was the most prosperous, the most lively and the most congested of all the camps.

'Through the straggling purlieus of the place we trot with crack of whip and warning shout', wrote an early visitor. 'The roadway swarms with naked Kaffir and brawny white man. Dressed in corduroy or shoddy, high-booted, bare as to arms and breast, with beard of any length upon their chins, girt with a butcher's knife on belt of leather—one could not readily believe that amongst these bronzed fellows might be found creditable representatives of every profession. The roadway grows white. Our wheels sink in "diamondiferous sand". . . . Thicker and thicker stand the tents, closer presses the throng. A din of shouts and laughter fills the air. We pass large drinking shops full of people; negroes go by in merry gangs. One stands amazed at such a crush of dwellings, such a busy, noisy host.'

The carts and tents of the early diggers at New Rush were slowly replaced by wood and iron houses. Between this camp and De Beers a straggle of streets developed into a town. The town was named Kimberley in honour of a British Colonial Secretary who could not pronounce the name of the farm, Vooruitzicht. As Kimberley it changed the course of a nation's history. In this dust-blown shanty town were founded the fortunes of many men.

PART ONE
THE DIGGERS

CHAPTER ONE

THE SETTLER'S SON

WHAT the Normans were to Britain, and the Pilgrim Fathers to America, the 1820 Settlers have become to English-speaking South Africa.

The 1820 Settlers came from Britain. They were brought out to strengthen the eastern frontier of the Cape Colony. In those days this frontier formed the boundary between white and black South Africa and, as was to be expected, conditions in the eastern Cape were precarious. Cattle thefts and disputed territory created friction among the races. The whites were outnumbered by their Xhosa neighbours and, to rectify the position, the Governor of the Cape persuaded the British Government to recruit settlers for the sparsely populated area. The promise of land and a new life was attractive to Britons suffering from the depression which followed the Napoleonic wars; but it required courage to face the uncertainties of this perilous country. The settlers—some five thousand of them—came, struggled and triumphed. They had good reason to be proud of their achievements: achievements which created a South African legend. To say a man is of 1820 Settler stock is to vouch for his virtue. When a man makes such a claim for himself, he is expected to substantiate it: not everyone can.

Joseph Benjamin Robinson, one of the most powerful South Africans of his day, was proud of his Settler background. He claimed that his parents were English yeomen and were among the first to emigrate to South Africa in 1820. Like so many claims made by the controversial J. B. Robinson, this was to become a matter of contention. In one of the few studies made of his career, Paul H. Emden dismisses most of Robinson's pretensions and ends by saying: 'nor is the name of his family among the settlers who in 1820 emigrated to South Africa'. This is thrown in as the final insult: it is quite mistaken. Whatever doubts may exist about J. B. Robinson, there is no denying his origins.

The twenty-seven-year-old Robert Robinson, his wife Martha, and five children arrived at Algoa Bay in the Cape Colony on 15 May 1820. They were among a party of settlers, recruited in Surrey, who sailed from London in the brig *Brilliant* and their arrival is one of the most graphically documented episodes in the saga of the 1820 Settlers. Sailing in the same ship was another party from Scotland, supervised by the crippled poet Thomas Pringle. In his book, *Narrative of a Residence in South Africa*, Pringle has described the mixed feelings with which the British settlers approached their new country. 'Seated on the poop of the vessel,' he says, 'I gazed alternately on that solitary shore, and on the bands of emigrants who now crowded the deck or leaned along the gangway; some silently musing, like myself, on the scene before us; others conversing in separate groups, and pointing with eager gestures to the country they had come so far to inhabit. Sick of the wearisome monotony of the long sea voyage . . . all were highly exhilarated by the prospect of speedily disembarking; but the sublimely stern aspect of the country, so different from the rich tameness of ordinary English scenery, seemed to strike many . . . with a degree of awe approaching to consternation.'

Their consternation was fully justified. The glowing pictures painted in England quickly faded once they travelled up-country. Drought, pestilence, cattle raids by marauding tribesmen and the lack of military protection brought many of them near to destitution and some of them to death. The Robinson family was plunged into the misery of these early years. They settled in the Cradock district of the Cape, where they battled with the elements and lived in fear of sudden attack. This struggle for existence did not make for a placid family life and it was not until twenty years later that their sixth surviving child, Joseph Benjamin, was born on 3 August 1840.

By this time the worst struggles were over. The frontier districts were still far from peaceful, but the settlers had established themselves and were beginning to reap their hard-earned rewards. When Joseph Benjamin was six years old, fighting was renewed with the Xhosas and it may have been during this disturbed period that he first tasted fear. Years later he was to recall an occasion when, during their father's absence, his brothers, urged on by their mother, successfully defended the farm against a Xhosa attack. Too young to play a part in the defence, he nevertheless remembered the incident in detail—including the smoke from ransacked houses on the horizon. Since such attacks were then becoming less frequent, their effect was all the more traumatic. The very unexpectedness of the Xhosa raids served to toughen youngsters reared in the lonely outposts of the Cape. Joseph Benjamin grew into an exceptionally tough young man.

A school was established at Cradock during the early years of settlement, but neither Joseph Benjamin nor his brothers appear to have attended it. Instead, they were taught by itinerant schoolmasters who toured the farms and provided the settlers' children with a slip-shod education. Little is known of Joseph Benjamin's schooling: few records were kept and he performed no outstanding feats of learning. His claim to recognition, during these early years, lay in a very different direction. He was strong, he was obstinate. The local children feared him; it was his boast that, by the time he was sixteen, he had fought and beaten every boy in the neighbourhood.

Pugnacious, quick tempered and independent, he possessed remarkable self-confidence, even as a young man. He was still in his early twenties when he left home and set himself up as a wool merchant, first at Dordrecht in the Cape, then in the wilds of the Orange Free State. Few young men would have chosen such an unpromising occupation. The conservative, God-fearing Boers of the Free State did not take easily to strangers—particularly English-speaking strangers—and to win their trust was, in itself, a full time job. This seems not to have bothered young Robinson. He spoke the language of the Boers fluently and no doubt felt a kinship with their dour, fiercely-held desire for solitude. He travelled from farm to farm, sipping coffee and bargaining with the bearded, pipe-smoking men on the stoeps and accepting fat-impregnated cakes from their self-effacing wives. The Boers could have had few misgivings about this intense, humourless young man. A gangling six-foot, with piercing blue eyes and a tight-lipped expression, he spoke plainly, drank little, and gave the impression of stolid reliability. Not only was he accepted, he soon became welcome.

He covered hundreds of miles in his ox-wagon in his search for business. At first he brought wool and farm produce, but before long he was buying farms and breeding horses. The trading store that he ran in the little town of Bethulie is said to have been a model of its kind.

Aggressive as ever, he did not hesitate to join a Boer commando

when the Free State went to war with the Basutos in 1865. He was
just twenty-five, with little experience of warfare, but he quickly
assumed command of his contingent and featured prominently in
the unsuccessful attack which the Boers launched against the
Basutos in their mountain fortress. He emerged from the battle
unwounded but not unscathed: in later years he was to claim that the
rigours of this campaign were responsible for his increasing deafness.

   Two years after this clash with the Basutos, came the news of the
'Eureka' diamond. Robinson's business treks took him beyond the
Orange River and, like many others, he was alerted to the possibility
of stumbling upon river diamonds. He had, it seems, good reason for
keeping his eyes open. His trading concern, which had started so well,
had run into serious trouble and he was sorely in need of money. In
fact, it was later claimed that he had been—or was about to be—
declared bankrupt. 'I know of course that J.B.R. became insolvent
when he was a trader . . .', wrote an acquaintance, 'and that there was
the greatest indignation because of certain dealings prior to his being
sold up. There were also wide complaints when, having obtained
rehabilitation, he stood on his legal rights after he had made a lot of
money, and would not pay up the deficits, so he could not, or ought
not to have had any money when he started [at the diamond
diggings].'

   Desperate he might have been, but diamond prospecting took
time. Not until July 1868 did he meet with any luck. The account of
how he found his first diamond became one of his favourite stories.

   While waiting for his Malay driver to negotiate a river crossing,
he noticed a mass of brightly coloured pebbles. He picked some up
and put them in his pocket. Once across the river he drove on to a
small village where he intended to spend the night. Here he was told
of a diamond that had recently been discovered and sold for a large
price. The news disturbed him so much that he was unable to sleep
that night and the following morning he drove back to the river.

   He had no idea how to recognise a diamond if he found one. The
only common test he knew was that a diamond would cut glass; he
had also heard that a diamond rubbed against glass in the dark gave
off a phosphorescent glow. This did not help him in making a selection of the river pebbles and he was obliged to form a theory of his
own. He studied the pebbles closely and filled a bag with transparent,
angular stones. Then he drove on to a farm and begged a night's
lodging.

Once again his sleep was disturbed: this time intentionally. Retiring to his room after supper, he waited until he was sure that his Boer host was asleep, and started to experiment. He blew out the candle and, one by one, rubbed the stones from the bag against a tumbler. Nothing happened. All the stones in the bag seemed to be worthless. He flung himself on the bed in a pet. Then he remembered the stones he had put in his pocket the day before. After testing five of these stones he began to get bored with the whole business. The sixth proved worthless. The seventh made an unusual grating sound. He relit his candle and looked at the stone. It did not appear to be a diamond: it was dull and irregular in shape. But not knowing what a diamond should look like, he decided to continue with the experiment. After wiping the tumbler clean, he blew out the candle and started to rub. Suddenly the stone appeared to glow. He rubbed and rubbed again. The glow became distinct. Pressing a sharp side of the stone against the tumbler, he felt it cut deep into the glass.

Forgetting where he was he gave a shout of joy and began dancing round the room. When his startled host arrived to find out what was happening, Robinson explained that he had had a nightmare; it was, perhaps, an apt description of the first symptoms of diamond fever.

Strangely, Robinson did not go back to the spot where he had found the diamond. Instead he returned to his headquarters and then made for the Vaal River. Why he did this is not clear. He may have had doubts about prospectors' rights in the Orange River area, or it could be that he had heard that tribesmen were finding diamonds at the Vaal. He was later to claim that he was the first white man to search the Vaal for diamonds but this seems unlikely. At least two other speculators are known to have been in the Vaal River region at that time and it may well have been rumours of their activities that prompted Robinson to try his luck there.

After crossing the Vaal on horseback, he went to a nearby kraal and asked the Africans living there whether they had picked up 'any pretty stones'. They told him that two days earlier an old man, living some three or four miles away, had found a stone which might interest him. 'I gave a man half-a-crown to take me up to the hut to see this old man,' Robinson was to say. 'He must have been quite 80 years old. We began to parley. "You picked up a stone the other day?" I said. "Yes," and he crept into the hut and brought out a dirty linen bag, and took out some old rags enclosed in other

old rags, until at last he came to a 2 carat diamond. I offered him
£12 cash for it but he would not sell . . . I could not induce him to
take my gold so I said: "Suppose I give you goats for it? I will give
you 20 goats." This did the trick.' The goats cost Robinson 7s. 6d.
each and launched him on his career as a diamond buyer.

Robinson may not have been the first diamond speculator to
reach the Vaal, but he was undoubtedly one of the most successful.
He had a distinct advantage over his early rivals. A local Griqua
headman who was indebted to him for a past favour gave him
valuable assistance by directing him to a promising part of the river.
Here, helped by the Griquas and African labourers from his Free
State farm, he commenced full-scale operations. 'I brought up all
my boys, Basutos,' he said, 'showed them the diamonds which they
quickly began to detect. I made them surface search on the other
side of the river. In six weeks they had found about 30 diamonds
worth £10,000.'

He was helped by a young man of his own age, Maurice Marcus,
who had previously worked with him in the Orange Free State.
Marcus had emigrated to South Africa from London in 1862 and had
been introduced to Robinson by an uncle who lived in the Cape
Colony. The two young men struck up a business-like relationship
(one hesitates to say they were friends: 'I always found him honest
and straight,' said Robinson, 'and soon I invited him to accompany
me on my trading trips around the country') and shortly before
Robinson left for the Vaal they had entered into a form of partnership. As soon as Robinson had begun to find diamonds, he sent for
Marcus and their partnership was consolidated. However, Robinson
appears to have retained most of the say in the business: Maurice
Marcus, who died a millionaire, was never more than a shadowy
figure in his partner's hectic career.

Robinson quickly learned all there was to know about surface
searching for river diamonds. In no time he had acquired extensive
properties on both sides of the river, from which he recovered
diamonds in large quantities. By the time serious digging began—
after the discovery of 'The Star of South Africa'—J. B. Robinson
was a man to be reckoned with on the banks of the Vaal.

Socially he was not a success. The diggers tended to shun him:
they found him abstemious, taciturn and tight-fisted. Many unpleasant stories were to circulate about Robinson's relationship with
the river diggers. He was accused of cheating them in business

transactions and of bribing their African labourers to sell him diamonds. Most of these stories seem to owe more to Robinson's lack of popularity than to truth but they were to plague him throughout his life.

Unlovable as he was, his influence could not be ignored. The Free State Government, glad to have a friend at court, appointed him as a magistrate in the part of the territory over which they were trying to assert their authority. He was instructed to 'put down the rioting in the Free State Territory' but as that territory was the subject of constant dispute his jurisdiction was uncertain. Nevertheless, he seems to have played his part as a law officer with typical two-fisted assertiveness and to have gained the reputation of being 'pro-Boer'. His friendship with the Free State served him well: it enabled him to buy properties without any undue fuss. None of this increased his popularity with the diggers, but it certainly enlarged his fortune. As a land-owner, digger and diamond buyer, he had few equals on the Vaal River.

He became the first claim-owner to export diamonds direct to London. Most diamonds were sold to firms in Port Elizabeth and Cape Town, but Robinson found this unprofitable. Characteristically, he decided to by-pass the middle-men and start his own export business. The intial consignment was taken to England by Maurice Marcus. 'We arranged a red flannel belt around his waist, with a series of pouches, into which diamonds were placed,' said Robinson. 'Night and day he wore them until he got to London, preferring the pain to the risk of having them stolen.' For a long time afterwards, Marcus is said to have borne the marks of this uncomfortable belt.

[2]

Marcus was still in London when the 'New Rush' to Colesberg Koppie took place. Robinson was in the van of the rush. He was to claim, in fact, that he was responsible for the discovery of the dry diggings. According to a story he told, he stumbled on the Dorstfontein farm—where the diggings began—by accident. On his way to the Vaal River he had heard of an old lady who had picked up some stones at the back of her house. He went to investigate. 'It was very early in the morning', he said, 'and after a time this woman brought out two bottles of pebbles among which I found six or eight diamonds. I gave her four sovereigns for the lot cash. That was the

starting point of old Dutoitspan.' Like so many similar stories told
by Robinson, the details of this account tended to vary with the
telling. There may have been some truth in it, but it is certainly
open to question.

But he was undoubtedly among the first to reach New Rush. He
visited the dry diggings weekly, buying up diamonds and claims.
Finally he decided to transfer his business to the new camp. He set
up shop in one of the early buildings—a single-storeyed wood and
white plaster affair—in the cluttered, dusty main street and advertised
himself as a 'Diamond Merchant . . . having entered into arrangements with London and Paris Houses . . . now prepared to give
the HIGHEST PRICES for all Descriptions of Diamonds and other
Stones.'

He was acknowledged as one of the most important diamond
buyers at New Rush. It was impossible not to recognise his person
or his position. His above average height, his cultivated military
stride, his cold-eyed stare and his white pith helmet, singled him
out as one of the outstanding personalities of the camp and earned
him the nickname of 'The Buccaneer'. He was, it is said, 'the very
picture of health and energy'; a man who could command respect
and inspire awe. Not the least of his virtues was his absolute
dependability. Diggers with diamonds to sell did not have to
comb the bars and canteens for him. 'He was virtually a teetotaller,
and certainly never was a frequenter of the saloons that were
overpoweringly in evidence throughout the camp in those days . . .',
says one of his customers, 'and he was invariably to be found in his
office during the usual hours of attendance.'

Studiously avoiding the raffish element at the diggings, he
established himself as a man of substance. If, at times, his aloofness
was offputting, this was not a bad fault: reliability, in the ambiguous
business world of the diamond fields, was valued higher than
familiarity. For all his shortcomings, Robinson was known to be a
diamond dealer of high repute. 'Mr J. B. Robinson', it was said at the
time, 'is known as a man of modest and retiring habits. He "does his
duty in that state of life which it hath pleased God to call him" like
a Christian and a gentleman. He gives the highest prices for diamonds, and hands over his cheque promptly and politely to all his
customers.' This was more than could be said for some.

Respectability entailed responsibility. At the river diggings he had
been a fist-wielding member of the Diggers Executive Committee;

now, in his new found dignity, he was called upon to represent his trade. He was on the committee of the Standholders and Licence-holders Association; he became chairman of the Diamond Buyers Committee; he was president of the Diamond Fields Chamber of Commerce and Mutual Protection Society. When the Governor of the Cape, Sir Henry Barkly, visited the diamond fields in 1872, Robinson was elected to the Reception Committee and put in charge of arrangements for the Governor's escort.

Throughout the Governor's visit, Robinson played his part of leading citizen to the full. Not only did he arrange for Sir Henry Barkly to be met, but accompanied him throughout his tour of the camp. The situation at New Rush was far from easy. Many of the diggers were dissatisfied with conditions on the diamond fields and blamed the British authorities, who had recently annexed the territory, for their plight. Fears had been expressed for Sir Henry Barkly's safety during this visit. Robinson was to claim that he was largely responsible for turning the tide in the Governor's favour. 'I had a good deal to do with his reception', he said, 'and have a vivid recollection of all that occurred . . . I remember just before the tiffin on that occasion that Sir Henry Barkly asked me what reception he would probably meet with. I replied that all would depend on his speech. Previous to the tiffin Sir Henry Barkly was the most unpopular man in the country; half an hour afterwards he was the most popular.'

Sir Henry Barkly's visit was a high spot in Robinson's early career. He was thirty-two; a successful businessman, a senior citizen and an influential political adviser. Of his dignity and standing there could be no doubt. He might not have been a universal favourite, but he had no doubts about his own importance. Then, unfortunately, he had to spoil it all. A sudden display of his ungovernable temper produced a sizeable crack in the pompous façade.

[3]

New Rush was full of men of many talents. Doctors, lawyers and journalists doubled as traders, auctioneers, claim-holders and diamond dealers. In the tents and shanties of the ramshackle town, men conducted a variety of businesses. One of the more colourful characters of the camp was a rotund, dapper little dentist named

Ernest Moses. In addition to extracting teeth, Moses wrote an occasional column for a local newspaper and ran an amateur diamond buying business. As a dentist his services were unremarkable, his newspaper column was generally vitriolic and his reputation as a diamond buyer dubious. He had done many unwise things. His greatest mistake was in offending J. B. Robinson.

On the morning of 30 September 1872, Ernest Moses was strolling along the dusty main street of New Rush when suddenly the door of J. B. Robinson's office flew open and the proprietor rushed into the street flourishing a horse-whip. Grabbing Moses by the collar he began to beat the little dentist unmercifully. 'The pliant whip', it was reported, 'came down one, two, three, fitting the plump shoulders of Mr Moses as though it had been made for them . . . Mr Moses took the one, two, three, and then tried to run off, but Mr Robinson is swift of foot, as well as smart of hand and strong of arm, and so he said, "No, no, Moses, it will not do to run," and down came the pliant whip again, one, two, three, across the Mosaic buttocks. . . . By this time the neighbours had turned out to see the fun, and the head of the Postmaster-General had popped out of the little wooden window hole, and his specs danced with glee on top of his nose. The watchmaker across the way, and the people under the verandah of the Blue Posts were called to the fun. Dr Mathews, always on the spot when wanted, waited with his foot on the step of his cart, fearing to go lest he might be wanted; but seeing Mr Moses walk off, more frightened than hurt, took his departure, and the other people who saw the fun went home too.'

The incident caused a stir. It was thought that Robinson was angry with Moses because of the remarks the dentist had made about him during Sir Henry Barkly's recent tour. Moses went about threatening to sue his attacker. The case was scheduled to be heard at the magistrate's office but, on the day of the hearing, the large crowd that had gathered were told that the matter had been settled by an exchange of letters. Moses wrote to Robinson apologising for 'having lately used language to third parties concerning you calculated to cause you annoyance and vexation' and Robinson replied 'I, on my part, now beg to express my regret at having used my horse-whip towards you.'

It later transpired that the horsewhipping had been the result of more than slander. Robinson appears to have had good reason for suspecting that Moses had palmed him off with a doctored diamond.

He claimed that he bought a white 'river-diamond' from Moses and later discovered it to be a yellow stone worth much less. The stone had been burnt white. According to another diamond buyer, Joel Myers, Moses had confided that 'the secret of burning yellow stones white was only known to dentists'.

Moses naturally denied the charge. However, he was unable to prove his innocence. When he went to tackle Myers on the subject he received another hiding. 'In the course of Mr Myers' expostulations,' it was said, 'Mr Moses made out Mr Myers to be a liar and that of all things Mr Myers could not stand, and he, to show he could not, slapped his accuser's face, and kicked him out of the office, and threw his umbrella after him.'

Robinson had no cause to regret his part in the affair. In fact, he was publicly congratulated for having exposed a clever swindle. When, a few months later, he left on his first trip to London, his standing was as high as ever. 'Mr J. B. Robinson . . .,' reported the *Cape Argus*, 'has been at the head of the most extensive, enterprising and successful diamond buying firm almost from the commencement of the finds. Mr Robinson is colonial-born and this is his first trip to the old country; and all who know him wish him very thoroughly the fullest enjoyment of the success which his enterprise and luck have won for him.'

Nevertheless, trivial as the horsewhipping of Ernest Moses was, it was not without significance. It revealed an eccentric side of Robinson's character. Not everyone at New Rush was favourably impressed by his display of righteous indignation. The diggers had their own standards of justice and the public horsewhipping of a man half one's size was not among them. For a man of Robinson's pretensions there were other ways of settling an argument, even in the rough and tumble of the diggings. If J. B. Robinson was a man to be reckoned with, so it seemed was his temper.

At New Rush the incident raised a little dust which quickly settled. There were plenty of other distractions. Not the least were the variety of entertainers who were flocking to the diggings to try their luck. One of these arrived a couple of weeks after the Robinson fracas. He opened at the St James's Hall—the iron hut which passed for a theatre at Dutoitspan—on 29 October 1872. He was billed as 'Signor Barnato, The Great Wizard'. He performed conjuring feats; his *pièce de résistance* being 'The Magic Picture Trick, as performed by Signor Barnato before the Royal Family.'

The show was an immediate success. Every performance was packed and many had to be turned away. After a series of dreary male quartets, the arrival of a sorcerer was an amusing novelty. If anybody knew, nobody seemed to care, that the exotic Signor Barnato was really Henry Isaacs whose nearest approach to royalty had been behind the bar of the 'King of Prussia' public house in Whitechapel.

CHAPTER TWO

A WIZARD AND HIS BROTHER

To be born a Jew in the East End of Victorian London was to start life at a distinct disadvantage. Not that the Jews of England were discriminated against to the same extent as their co-religionists in some European countries. There were no ghettos, as such, and as the century progressed many of the objectionable legal restrictions imposed on English Jews were lifted: in 1855 there was a Jewish Lord Mayor of London and three years later Baron Lionel de Rothschild won his long-fought battle to be admitted as a Member of Parliament. These formal dispensations, however, affected only a small section of the Jewish population; in the slums of London—where thousands of Jews eked out a beggarly existence—prejudice was as rife as poverty.

Nowhere was this more apparent than in the Whitechapel district: a maze of mean streets, dark alleyways, crumbling buildings and sleazy lodging-houses. Two rival exile groups, the Jews and the Irish, divided this district between them. Street brawls and gang warfare were the result of racial friction; criminal activities flourished; robbery, violence, drunkenness and prostitution became the natural accompaniments of a degraded way of life. Henry Mayhew, in his famous mid-century survey, *London Labour and the London Poor*, summed up the area as 'the most dangerous thieves' den' in London. One needed steel nerves and sharp wits to survive in Whitechapel.

Isaac Isaacs, a Jewish shop-keeper living on a corner of Petticoat Lane, taught his children how to look after themselves. He had two sons and three daughters, born and brought up in the two dingy rooms above his second-hand clothes shop. The boys, Henry and Barnett, were prepared for life in Whitechapel from the moment they could use their fists. Two nights of every week were devoted to boxing lessons. In the backyard, Mr Isaacs instructed his sons how to hit first and ask questions later. Often the boys were joined by their cousin, David Harris, who seems to have got the worst of these sparring bouts. Henry, the elder brother, was the most accomplished fighter, though little Barnett is said to have packed a heavy punch.

Young David Harris was a far more serious type than the Isaacs boys. He attended Coxford's college and was soon working as a ledger clerk for a London export firm. Harry and Barney had little taste for so staid a career. They received a token education at the local Jews' Free School, leaving at the age of thirteen to help their father in the shop; their spare time was spent pushing an old-clothes barrow round a nearby market. Harry was the businessman, Barney the performer. It was Barney's quick-fire patter that brought the customers in, while Harry took the money. The combination worked well.

Of the two, Barney was undoubtedly the more engaging personality. Dynamic, cocky, with a sharp cockney wit and an eye for the main chance, he became a favourite with the hawkers in Petticoat Lane and the jewellers of Hatton Garden. He was irrepressible. Strutting along the backstreets of Whitechapel, whistling and swishing the canes which he sold for a penny, he divided his time between ogling the girls and making a quick shilling. He dabbled in everything that promised a profit: from fresh fruit to collar studs. Among his haunts were the side doors of the local Music Halls, where he would pick up tickets discarded by patrons during the first interval and sell them to anyone wanting to see the second half of the show. Once he was given two shillings for a ticket instead of the tuppence he had asked for. 'What did you do?' he was later asked. 'Do? Do? Why I never stopped running until I had put eight streets between me and the theatre.' The only person he respected was his brother who, in his own way, was every bit as astute.

Harry was two years older than Barney and tended to be more practical. He took an evening job as barman at the 'King of Prussia', a rowdy gin-palace owned by his sister Kate's husband, Joel Joel. Frequented by down-at-heel actors, dock labourers and tarts, the 'King of Prussia' offered ample opportunity for Harry to display his talents as a bruiser. On Saturday nights Barney would pitch up to lend a hand as chucker-out.

It was at the 'King of Prussia' that the brothers got to know the performers from the Cambridge Music Hall, where Barney quickly made himself popular as an unpaid scene-shifter. Harry's approach to the theatre was less philanthropic. As quick with his hands as he was with his fists, he decided to turn his conjuring talents to good use. While Barney—nursing ambitions to become a dramatic actor—memorised the speeches of his idol, Henry Irving, Harry perfected his sleight-of-hand. Dressed in cast-offs from his father's shop, he devised a knock-about juggling act and roped in Barney as his stooge. They began to appear (well down the bill) on the local music hall circuit.

At first, Harry, the comedian juggler, hogged both the act and the applause. Barney, with his red nose and baggy pants, was little more than his brother's assistant: appreciated more in actresses' dressing rooms than on the stage. Then, one night, a stage manager decided to intervene by shouting 'Barney too' while Harry was taking his bow. The shout stuck. 'Barney too' had a ring about it; it sounded more exotic than Isaacs. Harry tried out some variations—Barnyto and Barneto—and finally came up with Barnato. From then on they were billed as the 'Barnato Brothers'. Unfortunately, the change of name did not have the desired effect: they still found it difficult to get bookings.

It was their old sparring partner, David Harris, who first recognised the possibilities of South Africa. At the beginning of 1871, when the finds at Dutoitspan and Bulfontein were making the headlines, Harris called to see his cousins and announced that he was off to the Cape to look for diamonds. He produced a map and his steam-ship ticket and assured Harry and Barney that he intended making a fortune from the £150 lent to him by his mother. The brothers, impressed that their aunt was prepared to risk her life's savings on the venture, were nonetheless sceptical of David Harris's chances. They had good reason to be. Harris was not a good candidate for the competitive life of the diamond fields: he was far too reserved and gentlemanly. When he arrived at the diggings (having walked the 600 miles between Durban and Dutoitspan) he found himself out-classed by cut-throat speculators and his early experiences were far from encouraging.

However, he seems to have appreciated that his much sharper cousins would prove more adaptable. In his letters home, he played down his failures and emphasised the opportunities open to shrewd businessmen: not only in diamonds but in bartering of all kinds. Harry became interested. Whitechapel was no haven; neither the 'King of Prussia' nor the Barnato Brothers' act held much promise; his talents might well be better employed in South Africa. Then, to everyone's surprise, David Harris suddenly arrived back in London. His luck had changed. Playing roulette in a bar at the diggings, he had won £1,400 in an hour and had caught the next ship back to England to repay his mother's £150 and to do a little showing off. This proved too much for Harry. No doubt convinced that he could do even better than his cautious cousin, he set off for South Africa in the middle of August 1872.

[2]

Harry Barnato arrived at Cape Town on the R.M.S. *Danube* on Saturday, 7 September 1872. Here he came to a temporary halt. Having only just managed to scrape together enough money for the voyage, he could not afford the fare upcountry. He lost no time in remedying this. Within a week of his arrival, an advertisement in the *Cape Argus* announced that: 'Signor Barnato, the Greatest Wizard known, has arrived by R.M.S. *Danube* from Europe, and will give his astounding Performance of MAGIC (including his marvellous changes, startling effects, illusions and transformations) in the Mutual Hall on Monday Evening the 16th September, 1872.'

His first couple of performances were literally a wash out. A terrible storm, which lasted several days, kept people firmly indoors: only a handful turned up at the Mutual Hall. However, by the beginning of October the weather had calmed and the show he gave as 'The Great Barnato' in the Odd Fellows Hall was a huge success. He battered watches to pieces and tore up pocket handkerchiefs before returning them whole to members of the audience, a wave of his wand made bank clerks incapable of counting money correctly and a full glass of wine was plucked out of the hat of a startled dignitary. 'The different tricks performed,' it was reported, 'are too numerous to describe respectively. The "Marvellous Fishery" was in itself an evening's entertainment. . . . The performance concluded with the wonderfully mysterious "Decapitation" or "Living Head".' Harry had got off to a good start.

A couple of weeks later he opened at the St James's Hall, Dutoitspan.

'The cry is "Still They Come!",' it was announced when Harry switched his act from Dutoitspan to New Rush. The weather again proved treacherous; nevertheless, the diggers turned out. With the help of Harry Fredericks 'the London Comique Singer' and Fred

Montague 'the Topical Singer', Harry had a good run at New Rush. He played in another Mutual Hall—a battered galvanised iron shed on the Market-square—where he was forced to battle with the stifling heat and a boisterous audience. Often shouting down hecklers in mock Italian, he was able to rise above the sound of breaking bottles and noisy scuffles at the back of the hall. His stage presence was undoubtedly impressive. Many years later it was remembered that, in these early days, 'he appeared to advantage in a dress suit, for Harry Barnato then, though strikingly vulgar, was rather a goodlooking man'. Dark, with lively eyes and a handle-bar moustache, he looked every bit the romantic Italian he pretended to be.

A high spot was reached in November when 'Signor Barnato's excellent entertainment was honoured with the patronage of the nephew of his late Royal Highness Moshesh, Basuto chief. The black noble expressed his great satisfaction with the performance, and declared his intention of again paying the Signor a visit.'

A magician on the diamond fields could not escape suspicion. Harry, who was following the accepted custom and developing a few off-stage sidelines—including diamond buying—became very touchy on the subject of his wizardry. When a diamond theft occurred in November, someone wrote to a newspaper: 'If anybody wants to steal a parcel of diamonds, he must be as quick of hand as Signor Barnato.' This, Harry considered, was a bit near the knuckle. 'Fully appreciating the compliment intended to be conveyed,' he replied, 'I yet find it liable to misconstruction by some over-imaginative individuals who, in my capacity as a diamond buyer, fear that I might combine the attributes of the "Wizard". Allow me this opportunity of assuring them that I keep the two callings distinct, the only approach to "magic" in the former business being the astoundingly high prices I am always prepared to give for good stones.' The incident was treated as a joke, but the suspicion remained. Harry's diamond buying venture was not an immediate success. Even David Harris (who returned to the diggings in December) was wary of taking him on as a partner.

He continued at the Mutual Hall until the end of the year. On Boxing night he gave a 'Grand Christmas Festival' including 'His Great Feat of the *Aerial Suspension of a Human Being*, Sleeping in Mid Air! (Must be seen to be believed.)' But this marked the beginning of his theatrical decline: the final blow was the arrival of a Christy Minstrel troupe. Soon the Great Barnato was reduced to staging sparring exhibitions with an ex-policeman in the Market-square. This brought in a few shillings, but did not provide any luxuries.

He was obliged to move from his hotel room and live in a tent. This was not such a privation on the diggings, where the majority lived under canvas, but it was lower than Harry had intended to fall. He began to think seriously of borrowing his return fare to England. At last, David Harris took pity on him and found him a position with a diamond dealer in Dutoitspan. It was a bit of a come-down after his much trumpeted arrival, but at least he could eat regularly. It helped restore his morale and he gathered sufficient enthusiasm to write home urging Barney to join him.

[3]

Barney needed little prompting. With Harry's departure White-chapel seemed drabber than ever. Barney had taken over his brother's job as barman at the 'King of Prussia' (where he was a great favourite with his sister Kate and his three nephews, Isaac, Woolf and Solly Joel) and had attempted a solo act at the Cambridge Music Hall, but these were stop-gaps. He had no intention of becoming a publican and his conjuring was not up to Harry's standard. Everything seemed against his fulfilling his remaining ambition of becoming a great tragic actor. Despite the fact that he had memorised the entire role of Othello, no theatrical manager would take him seriously. A stocky five-foot-three, with fair hair and the face of a cheeky *gamin*, he did his best to look intense behind his metal-rimmed spectacles, but he was invariably greeted with hoots of laughter. His attempts at Hamlet's soliloquy were so unfortunate that the only way he could command an audience was by repeating it while walking on his hands. There seemed very little chance of his following in Henry Irving's footsteps.

With both Harry and David Harris in South Africa, Barney became restless. Nearly twenty-one (he was born on 5 July, 1852) he wanted more from life than the odd jobs he could pick up in Petticoat Lane. Kimberley, as New Rush was now called, seemed far more promising than Whitechapel. Harry's invitation confirmed his own intentions: he cut out smoking, saved up his fare and a little over and booked a passage on the R.M.S. *Anglian* which sailed to the Cape in July 1873.

His arrival at Cape Town on 5 August was less spectacular than Harry's had been a year earlier. For once he had no need to advertise himself. He had made more than enough money to get to Kimberley. Not only had he amassed nearly £30 in cash, but his brother-in-law, Joel Joel, had supplied him with forty boxes of dubious cigars with which, it was hoped, he would found an export business when he reached the diggings.

However, his short stay in Cape Town was not entirely uneventful. Years later he was fond of telling of an encounter at the Masonic Hotel the morning after his arrival. He was, he said, sitting on the hotel verandah when he was accosted by a prosperous looking man who asked him where he was going. When Barney told him, the man looked sceptical. 'Too late, boy, too late,' he said. 'Nothing left. I struck it rich, but the sands are dry now. Best take the next boat back.' It was fifteen years later, according to Barney, before he met the man again; their positions were then reversed; Barney was a well-known mining magnate and the man an envious speculator. 'How did you do it?' the man asked. 'By not taking your advice and going to Kimberley all the same,' replied Barney.

The story has to be taken with a pinch of salt. Barney Barnato could never resist a telling anecdote and would invent them more easily than he could remember them. However, this one has a semblance of truth. In August 1873, men were beginning to leave the diggings. The price of diamonds had begun to fall, many diggers believed that the diamondiferous soil was nearing exhaustion and confidence in the new industry was waning. None of this bothered Barney. For £5 he was allowed to load his luggage onto an ox-wagon travelling to Kimberley: he walked beside the wagon during the day and slept under it at night. The journey, which lasted nearly two months, was, he recalled, 'one of the jolliest times I ever had'.

By arriving in South Africa when he did, Barney just missed a young man who was to play an important part in his life: their ships, in fact, had crossed in mid-ocean. This quirk of fate has often been commented upon. But it is neither extraordinary nor significant. Had they both been in Kimberley at this time, it is doubtful whether they would have met. They moved in very different circles and had little in common. They would have made little impression on each other. For the burly, serious minded young man who had just left Kimberley for England had none of Barney's *panache*: he was gauche, unsociable and so badly dressed that, on the voyage from the Cape, his only pair of trousers came apart and had to be patched with canvas by the ship's sailmaker. This did not worry him. His mind was on other things. He was on his way to Oxford University. His name was Cecil John Rhodes.

CHAPTER THREE

A SICKLY YOUNG MAN

'WHY did I come to Africa?' Rhodes was to say. 'Well, they will tell you that I came on account of my health, or from a love of adventure, and to some extent that may be true, but the real fact is that I could no longer stand cold mutton.'

This was years later, when cynicism was expected of Cecil Rhodes. He revelled in the myths created by his admirers and liked to embellish them with a few witticisms of his own. Cold mutton was a borrowed symbol; it had nothing to do with his arrival in Africa. Nor, surprisingly, did diamonds. He arrived eighteen months after the discovery of 'The Star of South Africa' without, as far as one knows, giving a thought to the country's mineral potential.

He came to Africa—whatever he might say later—for his health; for his health and to join his brother Herbert who was trying to grow cotton in Natal. He was seventeen years old at the time, suffering from a defective lung and a weak heart. The family doctor had recommended that Victorian panacea—'a long sea voyage'. As Herbert was already in South Africa, the doctor's suggestion was easy to act upon. Cecil's family appear to have held out little hope of his making much of his life: young as he was, when he sailed for Africa there was no-one to see him off.

It is difficult to know what Cecil Rhodes was like as a boy. His early years are not well documented: one is forced to rely on the reminiscences of those who saw him through the haze of his subsequent reputation, or the equally distorted image created by his admirers. From these sources he emerges as a stereotype of 'the great man as a boy'; a dreamy, impractical youngster, living in a world of his own, not concerned with the common herd. 'My father,' he told J. G. McDonald, 'frequently, and I am now sure wisely, demolished many of my dreams as fantastical, but when I rebuilt them on more practical lines he was ready to listen again.' It is a pleasant picture. Many a despairing parent has drawn comfort from it.

Certainly Rhodes's father, the Rev. Francis William Rhodes, could have had little time for dreamers. He had been married twice, produced twelve children—nine sons and three daughters—and as the poorly paid vicar of Bishop's Stortford, Hertfordshire, he was in no position to indulge a whimsical son. By the time Cecil was born, on 5 July 1853 (he was a year to the day younger than Barney Barnato) two of the older boys had died, but even so it was a struggle to rear such a large family. Education alone was crippling. Of the seven remaining sons, two, Herbert and Frank, were sent to good public schools. Cecil arrived too late to enjoy a similar privilege. A lack of funds and his uncertain health made it necessary for him to attend the local grammar school; where, as a 'grubby little boy with ruffled hair', he did reasonably well in the classics but showed no aptitude for mathematics. He had an ungovernable temper and few friends. He may have been a dreamer, but this is not apparent from the few family letters that have survived. His brothers called him 'long-headed Cecil'.

One friend he did have was his aunt Sophy, his mother's sister. To her he confessed, after leaving school, that he wanted to become a lawyer or, failing that, a clergyman. Whether his father, who also wanted him to become a clergyman, could have afforded the necessary University education is not certain. Cecil's future was decided by his poor health. Soon after leaving school he fell ill and was shipped off to South Africa.

[2]

Unalike as they were, there are some interesting parallels in the early lives of Barney Barnato and Cecil Rhodes. Their birthdays fell on the same day, both of them appeared somewhat feckless compared with their elder brothers and it was the initiative of those brothers that brought them both to South Africa.

In some ways, Harry Barnato and Herbert Rhodes were of a type: restless, assertive, with an eye to the main chance and a more discernible degree of ambition. As a youngster, Herbert Rhodes even had a touch of the Barnato flamboyance. One of his tutors remembered him as 'a typical schoolboy—clever, volatile, with a face like indiarubber, and an extraordinary command of expression. He was a born actor. . . . He might have excelled in another calling, that of Blondin. When I have been out for a walk with the boys and we passed an unfinished house, he would run up the ladder and out on a horizontal pole, where, without apparent effort, he would stand unsupported haranguing his schoolfellows.' It was Herbert's self-assurance that had brought him to South Africa.

In the late 1860s, the Natal Land and Colonisation Company was trying to entice emigrants from Britain. The colony of Natal had been open to white settlers for little more than 25 years and was in urgent need of enterprising young men. The offer made by the Natal Land and Colonisation Company was attractive: a free grant of 50 acres, with the option to buy a further 100 acres for £120 payable over 12 years. Prospective settlers had to be familiar with farming and of good character. For country-bred men of sturdy ambition and few means it was a tempting proposition.

Herbert Rhodes had seized upon it. He had applied for a farm and was given a grant of land in the luxuriant Umkomaas Valley, a few miles from the colonial capital of Pietermaritzburg. Here, against the advice of older colonists, he had started his cotton-growing scheme. He cleared between 45 and 50 acres of euphorbia bush and planted them with cotton. The first crop—attacked by sub-tropical insects and strangled by weeds—was destined to fail, but Herbert's enthusiasm remained undaunted. His optimism determined Cecil's fate.

When Cecil—'a tall, lanky, anaemic, fair-haired boy'—docked at Durban on 1 September, 1870, his brother was not there to meet him. Like many another adventurous young man in Natal, Herbert Rhodes had gone to try his luck at the Vaal River diggings. However, he had arranged for Dr Sutherland, the Surveyor General of Natal, to look after Cecil until his return. Herbert's absence at this time was to give rise to one of the many legends concerning Cecil Rhodes's arrival in South Africa. He has been depicted as a 'solitary and forlorn' youngster, tossed upon an alien shore, friendless and unprepared for his new life. This is not quite true. He may have been relatively friendless but he certainly was not helpless. For all his alleged dreaminess, he had not left much to chance. With £2,000, lent to him by his aunt Sophy, he had more than enough to set himself up as a farmer and, a month before leaving England, he had taken the precaution of arranging for a letter of introduction to the Durban manager of the Natal Land and Colonisation Company. It was not for nothing that his brothers called him 'long-headed Cecil'—he rarely embarked on a venture unprepared. Even without Herbert, he was perfectly capable of looking after himself.

Nevertheless, he waited for Herbert. Dr and Mrs Sutherland made him welcome at their house, 'Gardenscliffe', on the outskirts of Pietermaritzburg. They found him a quiet, reserved young man, very different from his ebullient brother. Mrs Sutherland thought that he had outgrown his strength and looked with indulgence on his passion for reading. Her husband was quite convinced that he would end his days as a vicar of an English village. They seemed to find nothing exceptional about him.

Herbert is said to have found a few diamonds at the river diggings. But his prospecting was not sufficiently rewarding to lure him from cotton-growing. When he returned to Natal, he took Cecil to his plantation and within a few months they had 100 acres of cotton land under cultivation. The work was hard but Cecil enjoyed it. Throughout his life he was to reverence manual labour: he considered men who worked by the sweat of their brow to be 'decent chaps', the rest were loafers. 'Shouldn't do that,' he once told a friend who wanted to write, 'it is not a man's work—mere loafing. Every man should have active work in life.'

Active work certainly helped to restore his health. Despite the steamy heat of the Umkomaas Valley, he felt better in Natal than he had ever done in England. The Rhodes brothers lived simply. They had two huts on the farm: one they used for sleeping, the other served as a sitting-room-cum-store. Food was cooked for them by an African servant and when this became monotonous they would beg an occasional meal from a neighbour. On Sundays, as good sons of a parson, they would attend church in Pietermaritzburg. Occasionally Herbert would take the day off for a game of cricket in the nearby village of Richmond.

Cecil never went far from the farm. Whatever spare time he had was spent poring over the textbooks he had brought from England. He was still vague about his future profession, but he was determined to go to University.

His hankering after a higher education was shared by Henry Caesar Hawkins, son of a local magistrate. Together the two boys read the classics and discussed plans for getting to Oxford without outside assistance. It was probably this ambition that prompted Rhodes to make his first financial investment. Soon after he had settled on the plantation, he sent a small sum of money to Dr Sutherland who obtained shares for him in Natal's first railway: a two mile line then being constructed between Durban and its harbour. Unlike his brother, he wanted to insure against the possible failure of the cotton crop.

Herbert had none of Cecil's caution. He thought in terms of quick returns. Although his trip to the Vaal had not been particularly successful, it had unsettled him. When, a few months after his return, the rush away from the river to the dry diggings took place Herbert became even more restless. He waited until March 1871 and then decided to try his luck at Dutoitspan. More and more men were leaving Natal for the diamond fields. 'People out here do nothing but talk diamonds', Cecil wrote home. 'Everyone is diamond mad.' In June, his friend Henry Caesar Hawkins abandoned his books for the diggings.

Cecil remained on the farm. Not that he was unaffected by the diamond fever. Diggers returning to Natal had tempting tales to tell. 'To hear Rolleston talk and see his diamonds makes one's mouth water', he had written to his parents. 'Three whoppers, one worth £8,000, another £10,000 and another £9,000. The man who found the £10,000 diamond had offered his claim the evening before for 15s. and nobody would buy it!' But, exciting as all this was, he did not allow himself to be carried away. Diamonds were an 'awful enticement' but they were also a 'toss-up'; cotton, on the other hand, was a 'reality'. There was plenty to keep him occupied on the plantation. He had thirty Africans working for him and the cotton was due for harvest in July.

This second crop was of slightly better quality than that which Herbert had reaped a year earlier. On 25 May, Cecil exhibited a sample half-bale of cotton at the annual show of the Pietermaritzburg Agricultural Society and came near to winning the £5 that was offered for the best entry. He was very proud of this achievement. In later years, when anyone told him that something was impossible, he would reply: 'Ah, they told me I couldn't grow cotton.'

Eleven days after the Agricultural show, he entered into negotiations for another farm. Having sent a sample of his cotton home, he seemed all set to extend his plantations. He never did. Within a few weeks his plans were entirely changed.

It was the astonishing diamond finds at New Rush that changed the course of Rhodes's life. The announcement, in the middle of July 1871, of the diamond discoveries on Colesberg Koppie made even the most stable of men restless. Young Rhodes was no exception. Herbert Rhodes had been among the first to peg out claims at

New Rush. His luck was instant. In a list of early returns from Colesberg Koppie 'Mr Rhodes of Natal' is reported to have found '110 carats, including stones of 14, 16 and 28 carats'. This was enough to make even the doubtful Cecil sit up and take notice.

Even so, his methodical mind ruled his emotions. Not until the cotton had been harvested in October did he set out to join Herbert at the diggings. He left Natal in a Scotch cart drawn by oxen and took over a month to complete the 400-mile journey to the diamond fields. His luggage reflected his ambitions: alongside an assortment of digger's tools were stacked volumes of the classics and a Greek lexicon. He still regarded diamond prospecting as a chancy business. If cotton growing had lost some of its attractions, Oxford remained his ultimate goal.

[3]

New Rush was still very much a primitive camp when Cecil Rhodes arrived there. Most diggers lived in tents or covered wagons; a few corrugated iron sheds served as stores, canteens, hotels and billiard rooms and, although an attempt had been made at road planning, the general impression was of a 'great white canvas town'. Living conditions were bleak and sanitation practically non-existent. The heat, the flies, the bad drinking water and a lack of fresh vegetables created widespread illness. Diarrhoea, dysentery and enteric fever were rife. Many diggers suffered from a form of ophthalmia, caused by the vicious dust storms and the strain of sorting diamonds in the glaring sun. It was hardly the place to strengthen a young convalescent.

If Cecil Rhodes suffered from any of the prevalent diseases, he made no mention of them in his early letters home. His attention, like that of the entire community, was focused almost exclusively on the great hole that was diminishing Colesberg Koppie. By the end of November 1871, the diggings had reached a depth of some sixty feet and the weekly value of diamonds recovered at New Rush was estimated at between £40,000 and £50,000.

Already the mine was awe inspiring. 'Holding to one of the posts by which buckets are hauled up and down, you crane you neck over the edge, and look down into the gulf. You draw back in amaze, with an exclamation!' wrote a visitor, shortly after Rhodes's arrival. 'There is another world down yonder, sixty feet below! The crowd is almost as great as that around you. Naked blacks, diminished to the size of children, are shovelling, picking, and loading—hundreds of them in that cool, shadowed, subterranean world. They fill buckets with crumbling earth, and endlessly haul them up and down on pulleys. Some are swarming to the surface on rope ladders. There is an endless cry, and laugh, and ring of metal down below. Buckets rise and fall with the regularity of a machine. On top they are detached and emptied in a heap, ready for conveyance to the sieve. There are not many claims in the best part of New Rush where sorting can be done at the pit's mouth. The white, dry earth is carted off to the outer edge, and goes to swell the monstrous piles that lie there. Upon the surface—so much as is left of it, which is but roadways—what a swarm of busy men!'

Few were busier than Cecil Rhodes. The cost of claims had risen enormously in the few months that the mine had been open, but Herbert Rhodes had succeeded, after some speculation, in securing three valuable diggings. However, good as his position was, Herbert was far from satisfied. He was convinced that he had missed out on the more promising spots. It is said that early in the rush he had sold what appeared to be a barren claim and endured agonies when, shortly after the sale, he watched the new owner unearth a faultless thirty carat diamond. Easily deflated, the feckless Herbert had been considerably discouraged by this experience. He was soon itching for new excitements. Cecil's arrival had given him the excuse he was looking for. Leaving his young brother to supervise the claims, he went to inspect the Natal plantation and then sailed for England. Cecil Rhodes, eighteen years old and completely inexperienced, found himself in sole charge of the three diamond claims, each thirty-one-foot square and estimated at that time to be worth £5,000.

Rhodes tackled diamond mining with the thoroughness that he had applied to cotton growing. He surveyed the claims, weighed up his chances and estimated his profits. Soon he was writing home giving expert opinions on the value of diamonds. He found that the big yellow tinged ones were often deceptive: they had 'a nasty habit of suddenly splitting all over'. On the other hand, every stone was of some value: 'the great proportion are nothing but splints', he reported, '—but still of even these you very seldom find one that is not worth 5s.'.

Occasionally he would stumble on a valuable stone but, for the most part, it was a matter of hard work and perseverance. He was ready to apply both. Every day he spent hours scraping the earth on his sorting table with a flat piece of iron. When his African labourers failed to turn up (having made enough money to buy a gun and return to their kraals) he would strip off his coat and do his own digging. Although young, he appears to have had no illusions about the way money was made. Unlike his brother he did not pin his hopes on a spectacular find. 'Diamonds have only to continue at a fair price,' he told his mother, 'and I think Herbert's fortune is made.' But he was careful to point out that it would take at least four years to work out a single claim. In an average week he estimated that he found thirty carats and made about £100. This was more than many diggers realised but it was still a far cry from Herbert's dreams of overnight riches.

The difference in temperament between Cecil and Herbert Rhodes was not only apparent in their approach to fortune hunting: socially they were poles apart. Herbert had been a popular member of 'The Twelve Apostles'—a group of young men who shared one of the liveliest messes in the camp. Cecil took his brother's place in the mess, but was only nominally a member of the group. Lacking Herbert's gregariousness, he kept himself very much to himself.

Diggers were to remember the young Cecil Rhodes as a solitary figure 'with his hands deep in his jacket pockets, going silently and abstractedly to his breakfast' or 'in his shirt sleeves, seated on an upturned bucket, sorting with keen eyes the diamonds from the gravel on an improvised table in the open air, or reading a text book for his next examination at Oxford with one eye on his native workmen'. He always appeared much older than he was. When, in 1872, Herbert returned from England with their brother Frank, no-one would believe that Cecil was the youngest of the three.

One of the few friends he made was Charles Dunell Rudd. Like Rhodes, Rudd, a twenty-eight-year-old Englishman, had originally come to South Africa to recuperate after an illness. While Herbert was still in England, Cecil Rhodes and Charles Rudd had found themselves working adjoining claims and had decided to pool their resources. The partnership prospered. The two of them not only worked side by side in the diggings—often hauling their own 'pay dirt' to the sorting tables—but devised schemes for improving their capital: including the erection of an ice-making plant from which they supplied the thirsty diggers with cold drinks.

Rudd was one of the first men to succumb to that rare charm which Rhodes reserved for those he wished to win over. Having been educated at Cambridge, Rudd entered fully into a plan which would allow Rhodes to go to Oxford while he looked after their joint interests at the diggings.

However, before this plan could be put into effect, Rhodes suffered a serious set-back. In the winter of 1872, shortly after Herbert's return, he had a slight heart attack. This, the first murmuring of the disease that was to overshadow his life, seems to have been the result of overwork rather than of the unhealthy atmosphere at New Rush. As soon as he was sufficiently recovered, Herbert took him on a gold prospecting expedition in the Transvaal. It was hoped that the invigorating air of the high-veld would prove beneficial to the invalid.

Rumours of gold finds in the neighbouring Boer republic had long been circulating. Herbert, optimistic as ever, was determined to investigate them. The expedition proved fruitless as far as gold was concerned; but Cecil was sufficiently impressed by the Transvaal to purchase a 3,000-acre farm there. Herbert returned to the diamond diggings so bitten by the gold bug that he decided to sell his claims at New Rush and start off on another quest.

Seemingly restored to health and now a partner with Charles Rudd in a growing diamond concern, Cecil Rhodes felt that he was in a position to complete his education. He sailed for England at the end of July 1873—passing Barney Barnato at sea—and matriculated at Oxford on 13 October.

Within six months he was back in South Africa. During his second term at Oxford, he caught a severe chill while rowing on the Isis. The doctors, despairing of his recovery, advised him to return to a warmer climate. At twenty years of age he was given six months to live.

CHAPTER FOUR

THE BLACK FLAG REBELLION

SHORTLY before leaving Kimberley for his first visit to England, J. B. Robinson took part in an important ceremony. At the beginning of 1873, Richard Southey, the former Colonial Secretary, took up residence at the diamond diggings as Lieutenant-Governor of Griqualand West. The arrival of the new administrator was an occasion for celebration. Robinson, as owner of the finest equipage in the territory, lent his carriage and horses to carry Southey and his wife into town.

The official entry was a cheerful affair. 'His Excellency was expected at four in the afternoon . . .', it was reported. 'At three-thirty, the road was crowded. Every inch of it was covered with well mounted horsemen and well ladened vehicles, and by four o'clock, when His Excellency arrived, before or behind him was one living mass of people . . . the run into town was done at a splitting pace, the people cheering as they went.'

For the new Lieutenant-Governor of the recently annexed territory, the arrival in his future capital must have been both rewarding and forbidding. After a hot, dusty, seemingly endless journey across hundreds of miles of featureless countryside, the sight of the white tents and flashing roofs of the diamond diggings must have come as something of a relief. This initial reaction over, however, the impression of the town was hardly reassuring. From a treeless Market-square, the rutted streets splayed out at all angles. Most of the houses were of canvas or mud plastered stones; only the more important buildings—the shops, the offices, the hotels and the bars—were of corrugated iron. Even by late afternoon the heat was unbearable. Nor could the first glimpse of his future home, grandiloquently termed Government House, have given Southey much pleasure. It was simply a small, two windowed, corrugated iron cottage, with a wooden verandah in front and a straggly camel-thorn tree at the back.

But, *faute de mieux*, the Lieutenant-Governor and his sad-eyed lady were obliged to make the best of things. Impressive in plumed hat and lavishly embroidered uniform, Southey was sworn in immediately on arrival and that evening the couple were present at a firework display in the Market-square. The place was packed and it was claimed that the Lieutenant-Governor 'met with good wishes from everyone'.

This display of loyalty was short-lived. Richard Southey proved to be one of the most unpopular men on the diamond fields. By the time J. B. Robinson returned from his holiday in England it was evident how rapidly the new Lieutenant-Governor had fallen from favour.

The trouble that was brewing was as complex as it was serious. Discontent among the diggers—which had manifested itself when Sir Henry Barkly, the Governor of the Cape, had visited the diggings in 1872—was endemic and could be traced to many causes. The squabble among the various states and individuals laying claim to the diamondiferous territory had created hostile factions in the camps. Unscrupulous land speculators had exploited this hostility and worsened the situation. There was uncertainty as to how long the diamond industry would last; claims were being worked out, diamond prices were falling, import duties absorbed by the Cape Colony caused the cost of living to soar. The population at the diggings had begun to dwindle and taxation had increased. As the mines went deeper many claims were flooded and the authorities were blamed for not pumping out the water. Diggers felt that they were not sufficiently protected—either physically or financially— and that the administration of the diamond fields was inefficient if not, as some firmly believed, downright corrupt.

But perhaps the most inflammatory of the diggers' many resentments concerned the Africans. For with the discovery of diamonds, South Africa had inherited the industrial problems of the nineteenth century—the conflicts between capital and labour, the insecurities of the artisans and the unskilled workers, the rootlessness of the masses—and this, in terms of the country's racial composition, meant a sharpening of the division between the white and the black man.

Most diggers were only too grateful for the cheap labour force supplied by the tribesmen who flocked to the diggings, but they bitterly objected to the few Africans who were allowed to work their own claims. 'Ruin, financial ruin for the whites, moral ruin for the natives, these are the results of the attempt to elevate in one day the servant to an equality with his master . . .', thundered the diggers' newspaper the *Diamond Field*. 'Class legislation, restrictive laws and the holding in check of the native races, till by education they are fit to be our equals, is the only policy that finds favour here.' The so-called 'class warfare' of Europe was distorted into racial friction in South Africa. Feelings ran high. 'Nigger', a new term of abuse for the black man, was popularised on the diamond fields by American diggers. An illustration of the bitter resentment was when a digger asked to be excused from a jury because 'he hated the nigger' so much that he did not feel able 'to acquit him of anything'.

Africans were also thought to be at the root of the industry's ineradicable evil—I.D.B. (illicit diamond buying). This was the greatest single affliction suffered by honest diggers and, all too often, African labourers were responsible for smuggling stones to buyers of stolen diamonds. The need for regulations which would check I.D.B. and rid the diggings of 'vagrants' was foremost among the diggers' demands.

Richard Southey, an arch imperialist and a man of upright but rigid principles, was not the most suitable official to deal with the tense situation. Obstinate and conscious of his dignity, he tended to hold himself aloof and refuse to negotiate with the diggers' represen-tatives. Moreover, he made no bones about his sympathy with the Africans. 'The objects aimed at by the leaders of the opposition are . . .', he wrote to Sir Henry Barkly in August 1873, 'to exclude persons of Colour from the exercise of the franchise and to grant privilege to white persons purely because they are white . . . as until recently nearly all the land in the Province belonged to persons of Colour . . . there is great injustice in attempting to deprive them.' Laudable as these sentiments were, they did not increase the Lieutenant-Governor's popularity.

If Southey was intractable, his opponents were implacable. They were a curious crowd with mixed motives. 'Diggers from America and Australia, German speculators, Fenian head-centres, traders, saloon keepers, professional gamblers, barristers, ex-officers of the Army and Navy, younger sons of good family who have not taken to a profession or have been obliged to leave; a marvellous motley assemblage . . .' is how the historian, J. A. Froude, described the population of the diamond diggings when he visited South Africa at

this time. Most, if not all, of these oddly-assorted men were represented in the opposition to Southey.

Agitation against Southey's administration continued throughout
1874. The diggers formed a Committee of Public Safety and, at the
beginning of 1875, a fiery meeting was held at the Kimberley Hall.
After a number of strongly worded resolutions had been passed, the
cheering audience was called upon to arm itself and prepare to rise
in rebellion when a black flag was hoisted at the mine. The man
behind this revolutionary move was Alfred Aylward, an ex-Fenian
who had recently served a prison sentence for wounding a man in a
brawl. Vehemently anti-British and a brilliant orator, Aylward was
one of the most colourful characters of early Kimberley. With his
inclusion among the leaders of the protest movement events took a
more ominous turn.

More meetings were called, seven armed companies were formed,
and men paraded the streets shouldering arms. These militant
demonstrations were organised by a body calling itself the 'Diggers
Protective Association'. Southey took a firm stand. On 19 March,
1875, he issued a Proclamation warning all people against 'taking
illegal oaths or assembling in arms'. The following day the diggers
replied with a Proclamation of their own. They swore loyalty to
Queen Victoria and announced their intention to protect life and
property and maintain order. Southey sent an urgent appeal to Sir
Henry Barkly for troops.

[2]

It was not until things had reached this stage that J. B. Robinson
decided to intervene. As a prominent diamond buyer and a senior
citizen, he had done his best to remain aloof. He had little time for
Southey and every sympathy with the protesting diggers, but his
standing in the town made it impossible for him come out in open
defiance of the Lieutenant-Governor. Above all else, J. B. Robinson
respected established authority. Even now his intervention was of the
most tentative nature. He was approached by two diggers and asked
to arrange a meeting between Southey and the rebel leaders. This
he did. Unfortunately the meeting did not take place. As always
seemed to happen, the preliminary negotiations dissolved into a
wrangle about the composition and the nature of the proposed
meeting. Things came to a head in a more violent fashion.

At the beginning of April, William Cowie, a canteen keeper, was charged with unlawfully supplying twenty guns to Alfred Aylward. This gave the militants the pretext they had been waiting for. While Cowie's trial was still pending, the Diggers Association let it be known that any attempt to carry out a conviction would be resisted by force of arms. The case was heard in the Resident Magistrate's Court on the afternoon of 12 April. Cowie was found guilty and sentenced to pay a £50 fine or, alternatively, serve three months' hard labour. No doubt acting upon the advice of the Diggers Association, Cowie refused to pay the fine and was led from the dock to serve his sentence.

Immediately sentence was passed, spectators in the Court began to leave. 'It was apparent from the noise outside the Court that some movement was on foot,' reported the Acting Resident Magistrate. Some movement was indeed on foot. The black flag had been hoisted at the mine and armed men were rushing to the Court House. The police decided to act swiftly. Four Justices of the Peace were summoned to accompany the policemen detailed to escort Cowie to the gaol, some 250 yards away. As they approached the entrance of the prison, they were overtaken by armed diggers who blocked their path. At that moment police reinforcements, armed with rifles with fixed bayonets, appeared at the lower corner of the gaol. They took up a position in front of the rebels. It was a tense moment. There were some 3,000 people, armed and unarmed, swarming about the gaol; if either side opened fire, it would have resulted in a mass slaughter.

The situation was saved by the coolness of the Resident Magistrate who, despite a revolver being fired close to his head, agreed to take the rebel leaders to see Southey at the Council Chambers. Southey, stiff as ever, refused to grant an interview. However, the Acting Attorney-General agreed to release Cowie on being given a cheque for £50, not to be cashed until the sentence had been properly reviewed. With that the 'Black Flag Rebellion' fizzled out. Cowie was set free and the crowd dispersed.

Alfred Aylward was not there to see the end of his revolution. As soon as the back flag had been raised, he had bolted for the Transvaal, where, hearing that his plan had failed, he published his own death notice in a local newspaper.

J. B. Robinson had also missed the fun. On the afternoon of the rebellion he had been conveniently confined to bed by illness. He was quickly informed of what had happened. A group of business men hurried to his house and asked him to attend a meeting which

was being called to send a deputation to the Governor of the Cape.
Robinson claimed that he was not well enough to take any part in
the proceedings, but he let it be known that any move to consult the
Cape authorities had his blessing. This, he was to say, was his first
public involvement in the disturbances.

[3]

Meanwhile Kimberley was still far from settled. The Associationists refused to disarm and the Cape authorities were hesitant to send
troops to the diamond fields. Southey, in a heavy-handed attempt
to assert himself, made things worse by enrolling volunteers as
special constables. Then he took 'the insanely injudicious step of
beginning to arm natives'. Once again things became tense. Southey
renewed his pleas to Sir Henry Barkly for troops. He was backed by
a group of civic dignitaries who published a 'Protest' in the Government newspaper. They deplored the existing state of affairs and
demanded militant action from the Cape authorities. It was the publication of this 'Protest' that brought J. B. Robinson out into the open.
At a packed meeting in the Kimberley Hall on 5 May, Robinson
acted as chairman. The meeting had been called to consider the
'Protest'. Robinson opened the proceedings with a slashing attack
on Richard Southey. 'Hitherto', he said, 'I have abstained from taking any part in public matters. I think, however, the time has
arrived when a better state of things should be inaugurated. All now
ought to come forward. . . . It is much to be regretted that all our
endeavours were in vain and nothing would induce His Excellency
to throw off his official grandeur. Owing to his repeated refusals to
receive any deputations, the Associationists first enrolled, and then
the Government armed its volunteers, which was more deplorable
still . . . the Lieutenant-Governor committed a grave error in
forcing his subjects to take up arms. From the action of the Government the community is now thoroughly unsettled, and the moderate
men must come forward and place themselves between the two
parties. The conduct of the Government is incomprehensible, and
it is our duty in every way and by all means to deprecate their efforts
to crush the people.'
Robinson had an impressive platform manner. He spoke forcefully and could generate tremendous enthusiasm among his audience.
This, his political debut, was a huge success. The meeting resulted

in the birth of the Moderate Party with J. B. Robinson at its head. Although he announced that it was the intention of his party to mediate between the administration and the diggers, it was perfectly obvious from his initial speech where his sympathies lay. This was not lost on Richard Southey's supporters. As the mouthpiece of the Lieutenant-Governor, the *Diamond News* began sniping at the Moderate Party and its leader. It was scornful of the party's impartial pose and implied that, far from holding himself aloof, Robinson had been financing the rebels from the outset. Robinson heatedly denied the accusation.

However, it seems fairly certain that Robinson did his best to act as mediator. He contacted Sir Henry Barkly by telegram and tried to persuade the Associationists to disarm. His efforts were unsuccessful. The matter was settled by the arrival of troops at Kimberley on 30 June. There had been talk of the Associationists marching out to prevent the troops reaching the town but, as in the case of the Black Flag Rebellion, words proved louder than action. 'Troops arrived yesterday . . .', it was reported on 1 July. 'They paraded on the Market-square, and were much surprised at being pelted with oranges instead of bullets.'

At dawn the following day, five leading Associationists were arrested. This caused a considerable stir. As soon as it became known, there was a rush of sympathisers to the Court House; several leading citizens vied with each other to bail the prisoners out. J. B. Robinson was one of those who arrived too late to act as a guarantor. He gave his support in a no less positive manner. A few days later it was announced that a Defence Committee had been formed and that funds to provide the men with legal aid were being collected at Robinson's office.

The announcement provided the editor of the *Diamond News* with a heaven-sent opening. 'Our readers cannot fail to remember how shocked the Chairman of the Moderates was when we stated that money had been supplied to keep alive disaffection from a source little suspected . . .', he wrote. 'We observe now, however, that although he was so thoroughly shocked at the idea that he should so far forgotten himself as to have tipped in support of arms and armour, his office is now the head centre for collecting funds for the defence of the men charged with sedition and riot, whose armed movements he would not have backed with a penny piece. The Moderate leaders are the very men who are now the foremost in providing the sinews of war for the defence of the men whose acts they sent delegates to Sir Henry Barkly to save them from.'

It is not difficult to imagine Robinson's reaction to this renewed attack. Surprisingly, however, he did not reply. He seems to have been hoping that someone else would spring to his defence. Nobody did. He was left to contemplate his own revenge.

With the arrival of the troops, the last flickerings of revolt petered out. The Diggers Association was dissolved and an amnesty granted to those who had illegally acquired arms. Even the five men who had been arrested soon realised they had nothing to fear. So obvious was this that the truant Alfred Aylward decided to come to life again and claim his place in the dock as one of the ringleaders. The trial, when it took place, was a mere formality and ended in a whole-sale acquittal. At the beginning of August Richard Southey was relieved of his post as Lieutenant-Governor. This brought the unfortunate affair to a close. Grievances still had to be settled, but most people were content to let events take their course.

Most, but not all. Certainly not Mr J. B. Robinson. The prevailing conciliatory spirit was all very well, but a few scores were still out-standing. Not the least of these were the unanswered attacks that had been made upon him by the editor of the *Diamond News*. When it became clear that his supporters were not prepared to accept the challenge, he decided to do so himself.

[4]

A week after Southey had been relieved of his office, J. B. Robinson wrote a long letter to the *Diamond News*. 'I have waited patiently', he explained, 'for the time to arrive when a distinct denial should be given to the false and unscrupulous attacks made from time to time during the late disturbances . . . on that much maligned party called the "Moderates" of which I had the honour to be the Chairman.' What followed was not a defence of the Moderate Party but a detailed account of his own conduct. He attempted to show that from the time he had entered into the conflict he had acted without bias and done all he could to bring about a peaceful settlement. He drew attention to the threats made to his party by 'the Government organ', the *Diamond News*, and defied the local authorities to name any Moderate who had directly or indirectly assisted the Associa-tionists 'with the sinews of war'.

This, as far as it went, was all very well. The subjects he had
discussed had been purely political; he had every right to defend
himself. At the close of the letter, however, sweet reasonableness
deserted him. His tone became distinctly menacing. 'With regard to
the personalities indulged in by a certain writer in the *Diamond
News*,' he concluded, 'I can only say that his animosity towards me
personally is well-known—the particulars of which I may yet have
occasion to publish, or adopt such other means as I may think
proper to elucidate matters.'

The certain writer to whom he was referring was R. W. Murray,
the editor of the newspaper. Murray was a well-known Kimberley
personality. Then a man in his mid-fifties, he had been an early
pioneer of the river diggings and had started one of the first newspapers on the diamond fields. He was a born journalist and his love
of a newspaper fight was so great that at one time, it is said, he
edited two rival newspapers; attacking himself in the columns of one
and answering in the leading articles of the other. Robinson must
have known that his threat would be pounced upon.

It was. Murray answered the letter in the same issue. 'With the
threat on the concluding paragraph before us,' he wrote, 'we
naturally approach the subjects touched upon in Mr Robinson's
letter with great fear and trembling, because he tells us that it is his
intention to "elucidate" the malice of the writer in this paper. This
is, of course, a very dreadful threat coming from the quarter it does,
but notwithstanding the consequences which may follow, we shall
criticise the letter before us as freely and as fully as we should have
done had Mr J.B.R. left his concluding paragraph unwritten . . .
whenever Mr Robinson is in the humour for "elucidating" we shall
be quite ready to assist him.' He went on to tear Robinson's letter
apart.

Murray should have known better. Two years earlier he had
reported Robinson's assault on the unfortunate Ernest Moses. He
must have been fully aware that the one thing Robinson could not
stand was being made to look a fool. To adopt a sarcastic tone
towards J. B. Robinson was to invite trouble: it came sooner than
Murray could have expected.

Shortly after the newspaper had been published, some friends of
Murray's burst into his office with alarming news. No sooner had
Robinson read the reply to his letter than he had rushed to a local
doctor and borrowed 'the stoutest horsewhip he could get'. Return-

ing to his office, he had laid the whip on a table and demonstrated to callers exactly how he intended to deal with the impertinent editor. Not only did he threaten to horsewhip Murray but he announced his intention 'to do the thing effectually, and in the public street'. Murray was to claim that when first told of this, he had laughed it off. His friends had then pointed out that 'Mr Robinson was just the man not to strike a man of his own age, but to carry out his threat on anyone he could take advantage of'.

At mid-day Robinson was still reported to be brandishing the horsewhip in his office. The expected assault had become the talk of Kimberley. Murray decided to play for safety. Pleading that had he been a few years younger he would have adopted another course, he hurried to the Resident Magistrate's office and made an affidavit accusing Robinson of threatening a breach of the peace. A warrant was issued and Robinson was arrested on the verandah of his club where, it is said, he was 'indulging in that tall talk for which he has lately made himself conspicuous'.

There are conflicting accounts as to what happened at the Magistrate's Court. Murray, trying to avoid publicity, decided not to report the incident. The *Mining Gazette*, however, whose editor was a friend of Robinson's, gave it prominence under the heading: 'In Danger of His Life'. It claimed that Robinson had assured the magistrate that 'on considering Mr Murray's age and infirmities' he had no intention of carrying out his threat. This was heatedly denied by Murray the next day. Robinson had climbed down, he said, after he had been advised to do so and he had made a statement to this effect in court. 'Mr Robinson', said the *Diamond News*, 'has since he made some money on the Diamond Fields, been under the impression that he could ride the high horse over everyone who came between him and the wind. Mr Murray thought that this man of money should be taught that he must be a man of manners too.'

Whatever the truth, there was no doubt about the outcome. Robinson agreed to observe the peace, returned the horsewhip, and the case against him was dropped. A month later a notice appeared in the *Diamond News* in which the editor formally apologised to J. B. Robinson for any offence given by his editorial remarks.

The fracas did little to increase Robinson's popularity. However, his position as leader of the Moderates had undoubtedly given him a taste for politics. Earlier in the year, when the rioting was at its height, he had announced his intention of leaving South Africa and settling permanently in Europe. Now he abandoned this idea. He was undoubtedly rich enough to retire, but his political appetite had been whetted. He turned his not disinterested attention to the question of improving Kimberley's civic status.

CHAPTER FIVE

PUBLIC EXPOSURES

THE quelling of the insurrection on the diamond fields had repercussions other than the quarrel between J. B. Robinson and R. W. Murray. Not the least of these was the appointment of a Royal Commissioner to examine the financial position of the mining community and to recommend strict economies. The man chosen for this investigation was Colonel Crossman. He arrived at the diamond fields at the end of 1875 and opened his Court of Enquiry in the Kimberley Hall on 5 January 1876.

The result of the Crossman Commission was to diminish Government responsibility for the mining community, make company mining inevitable and put a stop to Southey's liberal policy towards the Africans. 'There was an end now', says Professor de Kiewiet, 'to the grog shops and the black prostitutes that had made a Kimberley Saturday night a scene to marvel at. Gone too were the native gun shops that encouraged stealing and the "swell niggers" who were the "go-betweens" of the illicit diamond dealers. But equally there was an end to the rights of natives as Southey had conceived them. They could no longer be permitted to hold claims or wash debris. In the interests of efficiency and economy natives on the diamond fields could henceforth only have one status—that of labourer.'

The sweeping changes that followed Colonel Crossman's investigations had a mixed reception from the diggers. Some of the reforms they had been agitating for were implemented but, on the other hand, many shady practices were stamped out. Not everyone welcomed the exposures that were made.

The first day of the enquiry brought its surprises. Most of the morning was spent listening to complaints from the diggers regarding the administration of the mines. Not until later in the day did Crossman get round to examining specific charges. One of these concerned the flooded claims and the inefficiency of the water-pumping system.

It was said that a Mr Heuteau, who had charge of the machinery for pumping water from the De Beers Mine, had been offered £300 by a speculator if he would stop the machinery by damaging it in some way. Crossman sent for Heuteau who was sworn in and agreed, under oath, that he had been offered a bribe. When asked to name the speculator concerned he refused. He said that he had given his word not to divulge the man's name. Crossman then threatened that if this information was withheld he would take legal action to obtain it. Eventually, Heuteau suggested a compromise: he would not make a public accusation but if he was given a piece of paper he would write the name down. He was allowed to do this and, on being provided with paper, he wrote—'Mr Cecil Rhodes'.

There seems to have been little point in Heuteau's attempt at secrecy. Immediately Crossman read the name, he asked for Rhodes to be sent for. Rhodes was nowhere to be found. Instead, his partner, Charles Rudd, arrived and told Crossman that he was prepared to give evidence that Heuteau had perjured himself. Cecil Rhodes, he declared, 'was the last man to attempt bribery'. Crossman had no alternative but to summon Rhodes and Heuteau to appear before him when the court met the following Friday. He let it be known that if Rhodes showed up he would be prepared to hear what he had to say any time that evening.

Rhodes arrived shortly after the court rose. He explained that he had been at Dutoitspan, where he already had water-pumping equipment installed. Emphatically denying Heuteau's statement, he claimed that the entire story was a fabrication. The following day he attended a special meeting of the De Beers Mining Board and appealed to members of the Board to assist him in clearing his name. The Board was not particularly helpful. The Chairman pointed out that it was simply one man's word against another's and that Heuteau had told him about the bribe 'some months ago'. In the circumstances, the Chairman said, there was little they could do and the case was best left to the Court of Enquiry.

There can be little doubt that Rhodes was worried. Since recovering from his bout of illness at Oxford, he and Rudd had done much to build up their mining interests. While others had been rioting, they had taken advantage of the uncertain state of affairs to extend their activities. In combination with two or three other young men, they had increased their holdings and looked about for ways in which they could augment their capital. The pressing need for an efficient water-pumping system had provided them with an excellent opportunity. By forming a partnership with an engineer named Alderson, they were able to put in a tender for pumping out the Dutoitspan Mine. The fact that the only water-pumping plant they were able to obtain was second-hand and badly in need of repair did not prevent their offer from being accepted. They immediately ordered another plant from the Cape and then set to work with the battered pumps which they hoped would last until the new equipment arrived.

It was very much a touch-and-go business. The pumps were constantly breaking down, fuel was scarce, and there were endless delays in bringing up the new equipment. Things were not helped by Rhodes's notorious absentmindedness. In later years Rudd was fond of telling the story of how the preoccupied Rhodes had almost ruined the new enterprise. Alderson was away and Rudd and Rhodes had been left in charge of the pumps. One night as Rudd was clearing debris at the edge of the crater, he glanced down and saw that Rhodes, who should have been working the engine, had left his post and was walking up and down abstractedly. Seconds later there was a loud explosion as the boiler burst—Rhodes had forgotten to supply it with water. They appear to have patched the plant up, for it was working sufficiently well for them to make a bid for a contract with De Beers Mine. There was still, however, no sign of the new machines arriving. Things had reached this stage when Heuteau brought his charge of bribery.

When the Court of Enquiry met on Friday, 7 January 1876, both Rhodes and Heuteau were present. Rhodes had taken legal advice and was able to prevent any further investigation by announcing that he intended to charge Heuteau with perjury and had handed the matter over to the public prosecutor. The affair attracted a good deal of comment: for the first time the activities of Cecil John Rhodes were open to public discussion. 'The Attorney General will, we hope, not shrink from his duty', wrote R. W. Murray in the *Diamond News*. 'The charge has been made; the character of a respectable citizen has been assailed. If Heuteau can prove that his allegation is true, the Attorney General ought to put the law in force against Rhodes; but if he cannot, then Rhodes ought to have full justice done him, for if a man be robbed of his good name, he has suffered an injury which, in some cases, it takes a lifetime to remedy.'

But it was not to be as simple as that. During his lifetime Rhodes was to feature in a number of legal proceedings and more often than not the outcome was to be left in doubt: this first occasion was to prove no exception.

The preliminary hearing took place in the Resident Magistrate's Court six days later. Witnesses were called by the prosecution to testify that Heuteau had told them before the Enquiry was held that Rhodes was not the man who had offered him the bribe. Heuteau, describing himself as a thirty-five-year-old engineer from Mauritius, reserved his defence. Some doubt was cast on the legality of the case as, with Colonel Crossman being the only person to see what Heuteau had written, there was no evidence before the court that the accused had named Rhodes. However, it was decided that a *prima facie* case had been made and Heuteau was committed for trial on a charge of perjury.

Then, without any reason being given, the case was dropped. At the next hearing the Attorney General announced that he would not prosecute and no more was heard of the bribery charge. Whether there was insufficient evidence to pursue the matter, or whether a legal technicality prevented the prosecution is not known. On the face of it, the case against Heuteau certainly seemed better founded than his accusation against Rhodes. Not only had witnesses come forward to deny the charge but, as was later revealed, Heuteau was in danger of losing his job as soon as Rhodes's machines were installed. On the other hand, the Attorney General, Sidney Shippard, was one of the few friends Rhodes had made in Kimberley and would no doubt have done all he could to clear Rhodes's name had there been any chance of success. Heuteau was probably not the only one who was relieved to have the matter settled so easily.

The affair did nothing to enhance Rhodes's reputation among the diggers. Writing from England, a few months later, he was to say: 'My character was so battered at the Diamond Fields that I like to preserve the few remnants.'

[2]

One thing is certain: Rhodes did not suffer financially as a result of the accusation. Not only did he finalise his contract with the Dutoitspan Mining Board, but the De Beers Mining Board agreed to allow him to operate their machines until his own pumps arrived. The only remaining problem was that of getting the new pumps to the diamond fields. It was no easy matter. The railway had not yet reached Kimberley and recent rains had made the roads practically impassable by ox-wagon. Rain had also made things worse in the water-logged mines. At the beginning of February, impatient members of the De Beers Mining Board demanded that Rhodes attend a meeting to explain why he was not fulfilling his contract.

This was probably the meeting that Rhodes attended with Henry Hawkins—the friend with whom he had studied on the Natal cotton plantation. 'I have never forgotten', said Hawkins, 'the way in which he, still quite a youth, handled that body of angry men and gained his point, and an extension of time.' But it was not entirely a matter of Rhodes's persuasive personality. A report of the meeting shows that he gave a guarantee to have pumps at the mine within thirty days or forfeit £100. This show of confidence worked; but it was very much a gamble. Having heard that a farmer in the Karoo (the semi-desert between Kimberley and Cape Town) had a pump which he used for tapping underground water, Rhodes had made up his mind to buy it. This was to be the real test of his persuasiveness.

He set off for the Karoo in a Cape-cart and arrived at the farm eight days later. The farmer was not only startled at his audacity, but quite adamant about not selling the machinery. He explained that the pump was doing valuable work on his farm; he had no intention of parting with it. Rhodes refused to be put off. Day after day he returned to the farm, exerting his charm and increasing his offer. Finally the farmer—Devenish—gave in. For the exhorbitant price of £1,000, Rhodes bought the pumping engine and paid another £120 to have it transported to Kimberley. Not the least remarkable aspect of the transaction was the fact that Devenish, who had never seen Rhodes before, agreed to accept his cheques for the entire amount— one of which was made out in pencil. This implicit trust, Rhodes was to say, gave him a renewed respect for the Afrikaner race. The deal must also have confirmed—if confirmation were needed—his life-long belief that 'every man has his price'.

At the cost of £1,120 Rhodes had avoided forfeiting £100. But, of course, there was more to it than that. By fulfilling his contract he had secured his monopoly of the water-pumping systems. This, in turn, took him out of the ranks of the small-time diggers. The Rudd-Rhodes partnership was now on a firm footing and Rhodes himself was poised for much bigger things.

First, however, there was his interrupted education to be considered. No sooner were the new pumps installed than he left for Oxford once again. He remained there until the end of the year, returning to Kimberley only during the Long Vacation. While he was studying he did not neglect his mining interests; he was in constant touch with Rudd and managed to buy some reasonably priced pumping engines which he shipped to South Africa. 'By all means', he wrote to Rudd, 'try and spare me for two years: you will find I shall be twice as good a speculator with a profession at my back.'

The period following Rhodes's enforced return to the diamond fields—after his first illness at Oxford—is usually regarded as important by his biographers. Undoubtedly it contributed much to the foundation of his fortunes. His deal with Mr Devenish is often related (and embroidered upon) to illustrate his pertinacity, his resource, and his ability to wear down opposition. The bribery charge, however, has never been mentioned. His biographers appear to have been unaware of it. But it is not without significance. The fact that the charge was unproven does not detract from its interest: it epitomises the uncertainty that was to accompany many of Rhodes's business and political involvements.

[3]

On Thursday, 1 June 1876, the Theatre Royal in Kimberley announced that a benefit performance for Mr B. Barnato would be held the following Saturday night. The play chosen for this occasion was *The Orange Girl* which had recently completed a three hundred night run in London. An added attraction was the reappearance of Signor Barnato, 'the Wizard of the South', who had agreed to come out of retirement for one night to ensure the success of his brother's benefit. 'Signor Barnato', it was said, 'will give his wonderful magical entertainment which three years ago deservedly gained for him unbounded popularity amongst a criticising public on the Diamond Fields.'

The idea of a special benefit for young Barney Barnato was widely supported in the press. He had been appearing in various roles at the Theatre Royal for over twelve months and had won a great number of admirers. It was thought that the combination of *The Orange Girl* and Signor Barnato would bring 'a large audience, while Mr Barnato by right of his past exertions deserves a bumper house'. Certainly Barney could do with all the support he could get. His first couple of years at the diggings had proved far from prosperous. Arriving as he did, when the diamond industry was facing its first real crisis, he had found it impossible to make any headway as a digger or a diamond buyer. Only by using his wits, his talents as an actor, and a ready eye for a bargain had he managed to survive at all.

Many amusing and colourful stories are told of Barney Barnato's early days in Kimberley. He is said to have created a number of sensations with his feats as a conjurer, a boxer, and a practical joker. It is tempting to repeat some of the more outlandish antics attributed to him; unfortunately, most of them are extremely doubtful. Originating largely from Barney himself, the stories have been so embellished over the years that it is impossible to distinguish fact from fiction. Nor is it always possible to accept the accounts of diggers who claimed to have known Barney in these early days: for, as the Barnato legend grew, Barney eclipsed Harry in the popular imagination and many anecdotes concerning the cheeky young Barnett Isaacs owe their inspiration to the reputation established by his older brother. If only a fraction of the tales told about Barney Barnato at this time were true, he would undoubtedly have made a bigger splash in the news-hungry mining town than he did. As it is, contemporary accounts make it abundantly clear that, apart from his performances at the Theatre Royal, he was considered far less a personality than the flamboyant Signor Barnato. Harry could always rate a headline in the local press; Barney was very much in the background.

One of the few people who knew Barney Barnato well when he first arrived at Kimberley was Lou Cohen. Regrettably, Cohen is not the most reliable of witnesses. His book of reminiscences, in which he tells of his early association with Barnato, was later the subject of a legal action; some of the things he wrote were disproved. However, there seems no reason to doubt his account entirely. As far as Barnato is concerned, it appears far more accurate than some of the stories that were spread by Barney himself. Indeed a great many of Cohen's anecdotes can be verified from contemporary sources: this is more than can be said for Barney's far-fetched tales.

Lou Cohen, who came from Liverpool and claimed to have been educated in Belgium, had arrived at the diamond diggings in 1872.

He first met Barney Barnato in a Kimberley canteen. Barney was then new to the diggings and, like Cohen, was scratching a living as a 'kopje-walloper'. It was a demoralising occupation. Most 'kopje-wallopers' were little more than small time opportunists who, without claims or much capital, toured the diamond sorting tables in the hopes of picking up (or fiddling) a bargain. Respectable diggers had little time for such scroungers. However, when Barney breezed into the Scarlet Bar, throwing his cloth cap onto the hat rack and plonking himself down between two of the customers who were sitting close to Cohen, he appeared as confident as he was brash.

'It rather interested me,' says Cohen, 'to see the way in which he beamed on everybody in general but nobody in particular, without taking the slightest notice of the frowns and muttered curses of the two foreigners he had separated. He was a strongly built young fellow, wore a pair of spectacles on his uninviting dust-stained face, and had the ugliest snub nose you could imagine, but as good a pair of large grey blue eyes as ever flashed through a pair of glasses.' Barney introduced himself after choking on his soup and bespattering everyone within radius. 'You'll excuse me,' he apologised to Cohen, 'but a fly fell on my nose.' It was the beginning of a somewhat erratic friendship.

They met again at the diggings a couple of days later and compared notes. Barney explained that he was not doing too well as a diamond buyer because his bad eyesight prevented him from detecting flaws in the stones he bought. 'When a spot bobs up my brother 'Enry kicks up such a blooming row,' he said. If report be true, it was not only flaws he failed to detect. A popular story of Barney Barnato's days as a 'kopje-walloper' concerns a deal he made with a gullible-looking Boer digger. Standing next to the digger's table, Barney is said to have noticed a sparkling stone in the discarded gravel. He picked it up and offered to buy it for £5—the only money he had with him. The digger appeared amused, but reluctant to sell. They haggled for a bit and at last the digger accepted the £5. Triumphant, Barney rushed to the nearest diamond buyer's office to resell the stone, only to find he had been palmed off with a worthless piece of crystal. He could do nothing about it: at no stage had the Boer so much as hinted that the stone was a diamond.

The story, when it leaked out, did little for Barney's reputation as a diamond buyer and it is surprising that Cohen agreed to his suggestion of a partnership. However, Cohen says that he was completely disarmed by Barney's honest blue eyes and kindly expression and the two of them teamed up.

They set up office in a corrugated iron shanty next to a canteen on the Dutoitspan road. It seemed a promising spot for a diamond buying concern: diggers on their way to and from Kimberley stopped at the canteen for drinks and often needed to trade in a diamond for ready cash. Cohen put up £60 of the intial capital; Barney contributed £30 and the forty boxes of cigars he had brought with him from Joel Joel's pub in London.

It cost them a guinea a day to rent the shanty. Cohen thought this a bit steep but Barney disagreed. 'If you can make two pounds a day out of it, it ain't dear for a guinea', he said, 'and you've an office in the bargain.' The office part of the bargain was taken over by Cohen who dealt with visiting customers—sealing and often spoiling a bargain with the revolting Joel cigars—while Barney spent the day kopje-walloping. For the first month or so they slept on the floor at the back of the office and ate a nightly meal of curry and rice by the light of two candles stuck in beer bottles. Eventually they were able to move into the canteen for bed and board. The business became a going concern.

Barney undoubtedly had the hardest job. While Cohen sat in the cool office, he sweated it out at the diggings. The heat, the flies, the dust and his bad eyesight made every day a hell. But he remained as jaunty as ever and was at his best when competing for stones at the sorting tables. His early training in Petticoat Lane stood him in good stead when it came to haggling with the diggers. Hard experience soon taught him to recognise a flawless stone when he saw one.

In time Barney was able to buy a pony and cart. This not only made life easier, but helped to extend the business. The pony was a remarkable animal. It had once belonged to a diamond buyer who had secret connections with a number of Boer diggers; this buyer was rarely seen at the sorting tables but his daily round of the diggers' tents was known to be as profitable as it was guarded. Barney, it is said, was the only person to notice that the docile pony driven by this man was so well trained that it came to a halt of its own accord at every stopping place on the daily round. When it was announced that the buyer was leaving the diggings and selling his equipment by auction, Barney quickly persuaded Cohen that it was worth paying £27 10s. for the seedy-looking pony. It proved a sound investment. Not only did the pony know which tents to visit but, says Cohen, it 'was introduction enough to get into conversation with the Boers, and we made much money out of Barney's inspiration'.

It was not all work. In the evening the partners would do the rounds of the local hotels and billiard rooms where Barney amazed Cohen with his social versatility. Joker, mimic and raconteur, he was a match for anyone at a billiard table and practically unbeatable as a domino player. What proved a greater asset in the Kimberley bars was Barney's ability to handle himself in a fight. Cohen came to rely on this ability both in their social and their business life. 'Barney Barnato', claimed the admiring Cohen, 'was the best sparrer in Kimberley.'

There were times, however, when Barney's resourcefulness was of little help to his cautious partner. On a visit to two coloured prostitutes who had established themselves in a house close to the synagogue, for instance, Cohen lost out badly. There were two rooms in the house, each lit by a candle, and Cohen found neither the girls nor the atmosphere particularly inviting. He left it to Barney to lead the way and start the bargaining. 'When I heard the words "Five pounds" muttered, I reckoned it was time to trot,' he says. 'But Barney hopped into the other apartment, and I had time to see him show the damsel a five pound note, when the candle went out.' Cohen waited for his partner in the street. He did not have to wait long. Within a matter of minutes Barney came rushing from the house, shouting to Cohen to run. Taken aback, Cohen began to argue. He saw no reason why he should run and called Barney a fool for paying the girl five pounds. 'Go on,' panted Barney, 'I showed her a five pound note, but had a piece of paper in the other hand, and when the candle went out I gave her that instead.' This, and the noise coming from the house was sufficient to make Cohen see reason. With the screams of the girls growing louder and lights appearing at the windows, he took to his heels and dashed after the fleeing Barnato.

One of Lou Cohen's proudest boasts was that it was his partnership with Barney that founded the Barnato fortunes. He liked to refer to their makeshift office as 'the cradle of the Barnato millions'. He claimed that it was only his objection to Barney's gambling that caused a rift in the partnership. His description of their final quarrel —with him piously pleading with Barney to mend his ways and finally taking over Barney's share in the office for £125 and buying up their 'stock of diamonds'—is not very convincing. If they did indeed have a stock of diamonds when they parted, there was little evidence of it in the months that followed. There is a more feasible explanation for the collapse of their short-lived partnership.

Like many other small diamond buyers, they were probably forced out of business by the depressed state of the diamond industry. For they had hardly got going before the rumblings that led to the Black Flag Rebellion made themselves heard. Diamond prices were falling, confidence in the diggings was on the wane. It was not the best time to start a new business. Significantly, Cohen makes light of the turbulent political situation and has little to say about the economic crisis. It is, after all, more romantic to pose as a millionaire's first sponsor than to admit sharing in a failure.

Whatever the cause of the break-up, they parted friends and continued to see a good deal of each other. Barney's movements are somewhat vague during the months that followed. He appears to have drifted: picking up odd jobs where he could, living mostly on his wits. He frequently visited his cousin David Harris—who was now established as a diamond buyer—and is said to have pulled off at least one profitable *coup* while looking after the business during Harris's absence.

A story that Barney was fond of telling against himself probably also belongs to this period. He was invited to dinner, with some other 'poor relations', by a well-known Jewish hostess. It was a grand affair. The table was laid with massive silver; a butler hovered in the background. During the meal, Barney noticed one of his fellow guests slipping a huge silver soup spoon down the side of his boot. He said nothing. After dinner he offered to entertain the guests with some conjuring tricks. The hostess was delighted and agreed to allow him to use one of the soup spoons from the table for his performance. After juggling with the spoon he made it disappear and then instructed the butler to 'look in that gentleman's boot over there'. The spoon was recovered with much applause. 'And', Barney would say, 'I went home with the other spoon.' How true this is one does not know: it was how Barney liked to present himself.

What is more certain is that from this time on he spent more time back-stage at Kimberley's theatre. The original galvanised iron shed, in which Harry Barnato had made his debut, had been replaced by a more substantial hall. Known as the Theatre Royal, it was far from regal and not much of a theatre. Like most permanent buildings in Kimberley, it was a one-storeyed corrugated iron affair;

it had no corridors, boxes or gallery, and a bar ran the length of one side of the building. Often, it is said, the noise from the bar was such that it tended to drown 'the heroine's lamentations or the denunciations of the villains'. The actors sweated behind paraffin footlights and singers struggled to keep up with the showy resident pianist who spent most of his time eyeing the chorus girls. For all that, the theatre was extremely popular. Visiting theatrical companies were always sure of good houses; some variety shows ran for weeks with scarcely a change of programme. Barney, stagestruck as ever, was fully alive to the theatre's potential. The chance to shine at the Theatre Royal was far greater than it had ever been at the Cambridge Music Hall.

He seems to have started as a stage manager: designing scenery, organising costumes and acting as property master. Before long, however, he was appearing on the boards. He was not easy to cast—his size, his youth and his appearance were against him—but in time he built up a repertoire of character parts which were constantly in demand. His performances in contemporary melodramas and farces —*The Ticket of Leave Man, Pride* and *The Two Orphans*—were highly praised. He was equally at home wringing his hands as Fagin in *Oliver Twist* or overacting Iago in *Othello*. But his greatest triumph was as Mathias in *The Bells*. He had learnt the part long before he arrived in South Africa. The role had been created in London by his idol, Henry Irving, and, having studied Irving's interpretation down to the last gesture, Barney was able to give a passable impersonation of the great actor. It made his own performance appear surprisingly professional. 'When we saw the piece', wrote a critic, 'we joined most heartily in the well merited applause which the really good acting of Mr Barnato elicited . . . Mr Barnato in delineating "Mathias" was as unlike an amateur as it is possible to imagine, and he must we feel certain have studied the character for months, otherwise he could not have rendered it as perfectly as he did. . . . After the first act the audience demanded three curtains and a special call for Mr Barnato.'

Not everyone was so enthusiastic. Lou Cohen, who dabbled in the theatre himself, claimed that Barney's Mathias was ruined by his appalling accent and a lack of sound effects. His dramatic "Ow the dogs do 'owl', says Cohen, was 'a monstrous inexactitude considering the animals were as absent as his h's.' But then Cohen was probably jealous of Barney's popularity. After bringing a show to an abrupt halt, by starting a fight with a doorman who would not allow him in without a ticket, Lou Cohen's own standing at the Theatre Royal was not particularly high.

[4]

It must have been at this time also that Barney moved in with his brother Harry. This in a way was an admission of defeat. There is no evidence that the brothers had had much to do with each other during Barney's first two years in Kimberley. Lou Cohen is probably right when he says that Barney lived in fear of Harry and for a long time did his best to avoid him. Only dire necessity seems to have brought the Barnatos together again.

Harry had fared slightly better than Barney. He had married, left the diamond buying business at Dutoitspan, and set himself up as proprietor of the London Hotel in Kimberley. This was not quite as impressive as it at first appears. Although advertisements for the hotel claimed '. . . be it from Cape Town, Port Elizabeth, Natal and other parts, the first words from the passengers are "Please direct me to the London Hotel kept by Signor Barnato" ', they did not specify who these enquiring passengers were. The fact is that the ramshackle London Hotel depended more on its notoriety as a gambling den than on its fame as a hostelry. It was not unlike the 'King of Prussia' in Whitechapel. A few theatrical families boarded there and Harry did his best to keep up a semblance of respectability; but nothing could disguise the seediness of the hotel or the questionable nature of its clientele. The police, it is said, took a disconcerting interest in both the hotel bar and its habitués. If a policeman showed up unexpectedly at the London, claimed Lou Cohen, 'some of the most eminent customers would scatter like rats that had seen a cat'.

Barney was given a small wooden bedroom at the back of the hotel and for a while was very much under Harry's thumb. Dates are very difficult to check at this period of the Barnato brothers' careers but it is claimed that about this time (1875) they started a moderately successful diamond buying business. No evidence is produced to support this claim and, if the family business was indeed founded when unrest at the diggings was at its height, it must have been very much a side-line for there is no mention of it in the local press, where even the smallest diamond dealers advertised.

However, modest as their initial enterprise was, they managed to accumulate enough money to buy four claims in the Kimberley

Mine at the beginning of the following year. These claims cost
£3,000 and, it is said, the brothers risked their entire capital to buy
them. They had no money left to buy mining equipment and if, as
so often happened, the claims were flooded or caved in they could
be wiped out overnight. It was a gamble which Barney persuaded
Harry to take. The combination of Barney's intuitive genius—and
the fact that the claims were in a promising sector of what had once
been Colesberg Koppie—was sufficient to convince Harry that with
luck and hard work they could make their first mining venture pay.
This, at least, is the story they told.

That the purchase of the claims was a great drain on the Barnato
finances is undoubtedly true. Neither Barney nor Harry appeared
very prosperous in 1876. In June, when Barney was given his benefit
at the Theatre Royal, Harry felt obliged to revive his neglected
conjuring act to ensure that the show was a financial success. Three
months later it was announced that: 'Signor Barnato anticipating a
professional tour through the whole of South Africa, and with the
advice and consent of his professional adviser, owing to the continued
indisposition of Mrs Barnato, has determined to leave this part for a
season.' In order to make this trip Harry had to auction off a lease on
the London Hotel and sell most of the furniture. He left for Bloem-
fontein in October. His holiday was to be paid for by performances
of 'his popular entertainment with which he has been so successful
on the Diamond Fields'. While there is no denying their poverty,
Harry's sudden departure on a theatrical tour hardly bears out the
claim that the brothers concentrated their energies on their claims
and sweated 'day and night' to make the venture pay.

Barney was left in charge of the business. He still appeared from
time to time at the Theatre Royal and occasionally enlivened the
staid meeting of the Diamond Dealers and Brokers Association. It
was very difficult for anyone to keep a straight face when young
Barney Barnato rose to speak and his irreverent interruptions often
made it impossible for the chairman to maintain order. Barney was
always good for a laugh; nobody seemed to notice that at times he
was doing his best to be perfectly serious.

CHAPTER SIX

THE STATUS SEEKER

LIFE in Kimberley became somewhat staid once the threat of rebellion disappeared. The mining community was essentially bourgeois and, with the political crisis at an end, the town quickly reverted to its humdrum ways. Newcomers to the diggings were constantly surprised at this placid atmosphere. To those used to the raffish mining camps in America and Australia, the lack of lawlessness on the diamond fields was almost impossible to credit. For instance, a visitor to New Rush in the early days had been astounded to find that diamonds were sent to Cape Town and Port Elizabeth by ordinary mail coach. A guard was not considered necessary. Only when a bag of diamonds was found to be missing (it was later discovered lying in the veld, untouched) did it occur to anyone that the coach might be robbed.

The upright South African colonists who pioneered the diamond fields had undoubtedly done much to set this respectable tone; the continuing uncertainty about the industry's future tended to discourage all but the tenacious, serious-minded digger. Try as they may, historians are hard put to breathe the spirit of the wild west into the stolid citizens of Griqualand West. Once the pioneering days were over diamond prospecting became a routine, rather dreary business.

Stories told by old South African diggers are tame compared with those of their counterparts on the Californian goldfields. There were, of course, a number of rogues and adventurers and a good deal of gambling, drinking, fighting and fornicating and this undoubtedly gave Kimberley a bad name. When set against other South African towns it was regarded as a centre of sin. 'All that is revolting in human nature may be found there,' declared one tight-lipped visitor. 'The libertines, the forgers, bird catchers, and outcasts of Europe found asylum there, as in Alsatia of old. . . . The vices of drinking, swearing, cursing, bullying, lying, cheating, and all kinds

of other utter abominations permeate society.' True as this was, it
was no worse than could be found in a hundred other crowded
industrial towns. For all Kimberley's brashness, the majority of its
citizens respected the law, even if they did not always abide by it.
Excessive violence—gun duels, lynch mobs and organised outlawry
—was practically unknown on the diamond fields.

Political and industrial demonstrations took place from time to
time and occasionally a private feud would erupt in public, but once
the Black Flag Rebellion collapsed, the threat of mob rule more or
less petered out. For the most part, Kimberley was distinguished by
its dusty, heat-drowsed lethargy. Any disturbance of the peace—
other than the inevitable drunken brawls—was regarded as a note-
worthy event.

Various religious denominations had lost no time in establishing
themselves in the town. Cheek by jowl with the bars, the brothels
and the gambling houses rose the churches. The Wesleyans, the
Anglicans, the Lutherans, the Roman Catholics and the Jews all
erected first canvas, then corrugated iron and finally brick places of
worship. The interiors, lit by paraffin lamps, might be stifling and
the backless benches uncomfortable but Sunday after Sunday the
town echoed to the sound of the faithful singing their lusty hymns
or chanting their fervent prayers.

Despite this generally law-abiding atmosphere, there were
occasional commotions. One such occurred early in 1876. At midday
on Saturday, 19 February, a crowd of angry men marched down
Main-street and milled around the office of J. B. Robinson, shaking
their fists, cat-calling and hissing. When the poker-faced proprietor
showed himself at the window, the shouting became so loud that he
was forced to duck out of view. Further down the street, another
group was gathered outside the office of Mr Selby Coryndon, a well-
known attorney. These men were in a more genial mood. They were
waiting for Anton Dunkelsbuher, a popular diamond merchant, who
was consulting his legal adviser before leaving town. When Dunkelsbuher eventually emerged from the office and signalled to the men,
they gave a loud cheer and rushed to lift him on their shoulders.
'But,' it was reported, 'Mr Dunkelsbuher, with great good taste,
requested them to desist, and so earnest was he in his entreaties that
his friends ultimately complied.'

The trouble had been caused by J. B. Robinson's notorious
temper. Sometime earlier he had bought a diamond from Anton

Dunkelsbuher for £640. He was then told that the diamond was rumoured to have been stolen and sold to the unsuspecting Dunkelsbuher by a middleman. There was, in fact, reason to think that the stone would feature in an I.D.B. case and that it would be impounded by the police. Knowing that Dunkelsbuher was about to leave for Europe, Robinson had demanded security as a safeguard against the diamond being confiscated. Dunkelsbuher, who was agent for one of the wealthiest diamond concerns in South Africa, had offered a letter of credit for £640 but his offer had been rejected.

Negotiations between Robinson's attorney, Mortimer Siddall, and Dunkelsbuher's attorney, Selby Coryndon, continued until the evening before Dunkelsbuher was due to leave. They ended in stalemate. Coryndon insisted that his client's security be accepted but this was not good enough for Robinson who flew into a rage and demanded the return of the money he had paid for the diamond. As Dunkelsbuher was as much a victim of the unfortunate transaction as was Robinson, he refused to make any cash settlement until the police proceedings were concluded. The upshot was that Robinson, seething with anger, applied for a writ the following morning to prevent Dunkelsbuher from leaving Kimberley. This is what had sparked off the demonstration by Dunkelsbuher's friends.

For such action to be taken against a highly respectable merchant like Anton Dunkelsbuher was bad enough; the way in which Robinson went about it was even worse. According to Selby Coryndon, the writ for Dunkelsbuher's arrest had been issued at nine o'clock on the day he was due to leave Kimberley. Hearing of this, Coryndon had pleaded with Robinson to settle the matter quietly in his office. Not only had Robinson refused, but he had deliberately delayed putting the writ into effect until noon, when Dunkelsbuher had already boarded the coach which was to take him to Port Elizabeth. The timing was unfortunate. Dunkelsbuher was a popular Kimberley personality and a large crowd had gathered to see him off: it was this crowd that had stormed Robinson's office.

Luckily, Selby Coryndon was able to arrange bail for Dunkelsbuher who rushed to Dutoitspan where he succeeded in overtaking the Port Elizabeth coach. Despite the controversy surrounding his departure from Kimberley, he was to become one of the big names in the diamond industry. In London, where he settled, he built up a world-wide diamond business and many years later his South African representative, Ernest Oppenheimer, established the Anglo-

American Corporation—the most influential mining organisation in southern Africa today.

The unpleasant affair did little to help Robinson's shaky reputation. The following week Mortimer Siddall made an effort to defend his client by publishing long letters of explanation in the local press. They were not particularly convincing. 'I can assure Mr Siddall', wrote one of the mob, 'that his letter has not changed my opinion in the least, nor has it had any influence whatsoever except to remind those of us who hooted at Mr Robinson outside his own premises on Saturday last of that person's scurvy treatment of a highly respected gentleman who was leaving for home.'

Nor did Robinson help matters by spreading the story that Dunkelsbuher had known that the disputed diamond had been stolen. To accuse a man of dishonesty as soon as his back was turned was not the way to win friends in Kimberley.

[2]

At one time Robinson would not have been unduly worried about the hostility he aroused. Although hot tempered and vindictive, he could afford to ignore the attacks made upon him. He was rich and he was powerful. His diamond buying business, one of the oldest and most prosperous in Kimberley, represented only the commercial side of his many enterprises. For years he had been buying up claims and operating them with considerable success. 'In 1872 I bought some claims in Kimberley', he was to say. 'A man came up to me one day, and said "I can buy half a claim off Major Dartnell; just you buy it for me, and I will work it and go shares in the profits." He set to work on the ground which turned out to be immensely wealthy. In a month I said it was quite enough for me and that he could go on working the ground himself.' How true this is one does not know. Few diggers would have recognised the parsimonious Robinson in the role of generous benefactor.

His reputation for meanness was as well-known as was his bad temper. So penny-pinching was he, in fact, that he got up before dawn every day to buy his food on the open market. His frugal domestic habits remained with him throughout his life. Years later, when he was a multi-millionaire living in a Cape mansion, he would measure out the daily ration of tea and sugar and adamantly refuse to increase it. Visitors to his house were often startled by his austere mode of living. 'A man of many economies [Robinson] employed no surplus butlers or parlour maids, and he had a mania for saving light', says one caller. 'So the great house, as one approached it, was always wrapped in darkness. Not until one rang did a solitary light go up, and then on the every-millionaire-his-own-parlour-maid-principle, Robinson himself would emerge to answer the bell.'

However, his account of his early claim buying (for which one only has his word) does reflect the way in which he could spot a promising claim and buy out impoverished diggers. Like others who had the foresight and the money to stand firm during the politico-economic crisis of the mid-seventies, he had been able to take advantage of the prevailing uncertainty to expand his activities. The J. B. Robinson mining company had grown to such an extent that even his most hostile critics were obliged to respect the influence he wielded.

Had money been his only concern, he would have ignored public opinion. It was his decision to play a more prominent political role that changed things. Now he was more than ever conscious of his position as one of Kimberley's leading citizens. The pompous, self-important air he had earlier assumed was still very much in evidence. He was undoubtedly very concerned about the damage done to his public standing by the Dunkelsbuher affair.

Nevertheless, he continued to involve himself in local politics. At a meeting held the following month to propose that Kimberley became a municipality, he played a prominent part. In June he announced that he would be standing for the Legislative Council. His nomination was supported by several well-known capitalists—including Rhodes's partner, Charles Rudd—and he received valuable backing from the local press. Even the *Diamond News*, which had always criticised him, urged its readers to 'vote for Robinson'. This was partly due to the fact that R. W. Murray (whom he had threatened to horsewhip) had left the paper and was now editing the *Diamond Field*, but mostly because the new editor of the *Diamond News* had no love for Robinson's opponents. The most popular man standing for election was undoubtedly Henry Tucker, the former chairman of the Diggers Protective Association. Tucker was a man of undoubted integrity and his championing of the diggers' cause had earned him a large following. But for those, like the editor of the *Diamond News*, whose loyalties lay with the Imperial authorities, Henry Tucker was regarded as one of the most dangerous men in

Kimberley. Even the choleric, whip-wielding Robinson was preferred to the former rebel leader.

Robinson entered the election with all flags flying. And he had a
great many flags. His immense wealth gave him a distinct advantage
over other candidates. His red, white and blue colours and rather
improbable slogan 'Robinson for Ever', were displayed throughout
the town. Wearing his familiar white pith helmet and doleful expression, he addressed several meetings, lecturing voters on the responsibilities of office and his eminent suitability to undertake them.

The campaign lasted three weeks. In the course of it Robinson
started legal proceedings against one of his rivals, a Dr Murphy, for
'vicious attacks' made upon his private character. This was no more
than most people expected from the cantankerous Robinson, and it
could have come as no surprise that the threatened legal action was
dropped once the election was over. However, this was not the last
that Dr William Murphy was to hear of J. B. Robinson. Not only
did the two men patch up their quarrel, but Dr Murphy was to earn
dubious distinction as Robinson's henchman. 'The Devil himself',
it was said, 'did not know what a baneful thing he invented when he
sent into this world spiteful Dr Murphy, who in his life has been
known to serve only two masters—Satan and J. B. Robinson.'

But neither Robinson's well financed campaign nor his attacks
upon his opponents made any difference to the election results.
Henry Tucker and his running mate headed the poll and J. B.
Robinson ended up a very poor third. 'Could bunting have carried
the day,' commiserated the *Diamond News*, 'Mr Robinson had been
certain of success.'

Robinson's defeat came as a blow to the Kimberley financiers.
They had backed him to act as their spokesman and had fully ex-
pected him to romp home. 'Shows one can never tell', commented
Cecil Rhodes from England.

Henry Tucker's victory was short lived. Not long after the
election he was arrested for buying diamonds without a licence.
The arrest shocked everyone. Tucker's honesty was unquestioned
and his offence was a technical one—his diamond buying licence
having just expired. That Henry Tucker should have fallen foul of
the new diamond buying regulations was ironic. His leadership of
the Diggers Protective Association had been inspired by a genuine
desire for the reform of the illicit diamond buying laws: now that
reforms were being implemented he was one of the first to be

arrested. Hostile magazines in England published cartoons of the unfortunate Tucker captioned 'Hoist with his own petard'. The man from whom he had bought the diamonds was Barney Barnato's ex-partner, Lou Cohen. Like the majority of diggers, Cohen was extremely distressed by the affair. 'When it became known', he wrote, 'that Mr Harry Tucker had committed an offence against the Diamond Law—even though a technical one—his political enemies pounced down on him, and he was at once arrested. . . . I did all that was possible to help him, but the trial was a foregone conclusion, and, to my intense sorrow, good-hearted, honourable Harry Tucker was sent to prison for nine months.' Even the defeated J. B. Robinson recognised that Tucker had been unfairly treated and expressed sympathy for him.

[3]

There could be no hope of Robinson replacing Tucker as a member of the Legislative Council. He had fared too badly at the polls. Instead, he had to content himself with lesser posts. When, a month after the election, a meeting was called to form a volunteer cavalry corps in Kimberley, Robinson took the chair. His early experience in the Basuto war ensured him a commission in the Kimberley Light Horse. This delighted his critics. From then on they made a point of addressing him by his full style and title—'Captain Joseph Benjamin Robinson, K.L.H.'—whenever he became too pompous. And, of course, he became more and more pompous.

He regularly attended meetings of the Kimberley Mining Board and was invariably chosen by his fellow diamond dealers to represent them. Although he was far from being the most popular diamond merchant in Kimberley, his colleagues had no hesitation in electing him to look after their interests. Few could get the better of J. B. Robinson in a business deal. Shrewd and tightfisted, he was notorious for driving a hard bargain and squeezing every penny out of the most trivial transaction. His only handicap as a negotiator was his increasing deafness. This seriously hampered him in debate. On one occasion, for instance, he startled a committee by arguing against a resolution and then moving an amendment which, in effect, amounted to the same thing. Later he had to confess that he had misheard the original proposal. His deafness undoubtedly contributed much to his bad temper. Frustrated by faulty hearing, he was inclined to shout when making a point and then abuse those who shouted back at him. Feelings often ran high at meetings attended by J. B. Robinson.

In addition to all his activities, he was still determined to play a part in local politics. There was no question of his accepting his electoral defeat as final. His determination to obtain office of some sort was, in many ways, remarkable. At this time it was the one thing that meant more to him than money. To further his political career he took over the *Independent* newspaper. This was the first of his many attempts to control the politics of the mining industry from over an editor's shoulder.

One of his first acts on taking over the paper was to make important staff changes. He put his book-keeper, Lionel Phillips—a young Jewish clerk, recently out from England—in charge of the printing works and appointed his attorney, Mortimer Siddall, as editor. Siddall probably owed his appointment to his knowledge of law and his willingness to allow his employer a free hand in the political columns: both these assets were very much in evidence in his conduct of the paper. 'I purchased the *Independent* newspaper', said Robinson, 'and though I was not the editor, I wrote for it at times. I took a prominent part in local politics.' Certainly there was no mistaking the bias or the force of the *Independent*'s policy once it was in Robinson's hands. Editorial comment was so strong that, within months, both Siddall and Robinson were being sued for libel. Of the cases that came to court, the most important was that brought by the new administrator of Griqualand West, Major Owen Lanyon.

The choice of Lanyon as Richard Southey's successor was unfortunate. A melancholy, reserved and autocratic man, he had many of Southey's faults and added to them by attempting to rule Kimberley as if it were a military barracks. Cecil Rhodes was later to sum up Lanyon's administrative abilities by accusing him of conducting 'business on the lines of a second-rate line regiment'. From the time of his arrival, Lanyon had made no secret of his distaste for the primitive living conditions at Kimberley and his impatience with local dignitaries had earned him many enemies. To make matters worse, before coming to South Africa he had served for several years in Jamaica and this, together with his swarthy complexion, gave rise to the rumour that he was a half-caste. There was little hope of Major Lanyon overcoming local prejudice. He was a sitting target for aspiring politicians and opposition newspapers. However, the attacks made upon him by the *Independent* went further than most. They culminated in a demand for his arrest on a charge of illicit diamond buying.

In an editorial published on 16 November, 1876, Siddall accused Lanyon of buying a diamond from a dealer without a licence. The circumstances of the sale were said to be similar to those which had landed Henry Tucker in gaol. Siddall maintained that Lanyon should be made to pay the same penalty. Lanyon was quick to defend himself. A few days later it was announced that a civil action had been initiated against J. B. Robinson as proprietor of the *Independent* and that Siddall, as author of the libel, was to be prosecuted on a criminal charge. The case became a South African *cause célèbre*.

Little hope was held out for Robinson or Siddall. In fact, they made matters worse by publishing a half-hearted apology. 'Neither of these gentlemen has previous experience of journalism', said a Cape paper, 'and it is possible that both of them may have occasion to regret the rash enterprise in which they are engaged. They profess to entertain a great respect for the Major. . . . One of them is lovingly eulogistic of the Major's many good qualities. But somehow the eulogy did not appear until the alleged libel had spread far and wide, and its consequences threatened to be anything but pleasant.'

Outsiders reckoned without the temper of Kimberley's indignant citizens. For once J. B. Robinson was considered to be on the side of the angels. By attacking Lanyon his newspaper had performed a double service: it had hit at the unpopular administrator and highlighted the injustice of Henry Tucker's imprisonment. No Kimberley jury was likely to convict a man who had conducted such a worthy crusade. This was made only too clear at Siddall's trial on 7 February 1877. The court was packed. The slight defence that was put up amounted to little more than a technical quibble. After the judge's adverse summing up, Siddall did not appear to have a leg to stand on. None of this made any difference to the jury: after a brief adjournment, they acquitted Siddall on every count. 'The verdict', it was reported, 'elicited bursts of applause from the crowded Court, which the judge vainly endeavoured to suppress. . . . Mr Siddall was seized by the crowd, and, amidst vociferous cheering, chaired across the Market-square. Great rejoicing among the respectable portion of the people here, who form the bulk of the community.'

Robinson was not so lucky. The civil action, in which he was sued for £10,000, was heard the following month without the benefit of a sympathetic jury. From the outset it was obvious that the case would go against him. After he had admitted that he had seen the contentious editorial before the paper was published and had made no attempt to suppress it, the judge ruled that there could be no defence, only argument to mitigate damages. Wisely Robinson capitulated. He acknowledged that the editorial was libellous and false and agreed to publish an apology in papers throughout South Africa. Much to his relief, Lanyon accepted this and decided not to press the case further.

The Lanyon libel case proved a mixed blessing to J. B. Robinson. There can be no doubt that by championing a popular cause, he had at last won public approval. But if it had been an astute political move, financially he had lost out badly. Although he had escaped paying damages, he had to meet the costs of both his own and Siddall's case; these, in Griqualand West, were heavy. To make matters worse, two more libel actions against the *Independent*—brought by private citizens—came into court within weeks of the Lanyon suit. In both cases damages were awarded against the newspaper. Admittedly these damages were not exorbitant but once again Robinson had to meet the costs. Running a newspaper was proving an expensive hobby. It unsettled both Robinson's miserly habits and his staff of novices. Siddall resigned as editor and Lionel Phillips requested a transfer from the printing works to the firmer ground of the J. B. Robinson mining company. Shortly after this, Robinson let it be known that he intended selling the newspaper.

Ostensibly he did sell the *Independent*. He sold it to Casper H. Hartley. At the same time he retained his control. The way in which the transfer was arranged is not without interest. It provides an insight into Robinson's business methods.

Robinson accepted a bond from Casper Hartley as payment for the *Independent*. For years this bond remained unredeemed: Hartley made no attempt to pay it off. So, although Hartley was, to all intents and purposes, the owner and editor of the newspaper, he remained tied to Robinson as surely as if he had been a hired hand. This suited Robinson very well. The *Independent* was still his mouthpiece but he was able to disclaim all responsibility for the paper. Time and again he was attacked for items appearing in the *Independent*; invariably his reply was one of outraged innocence. 'I am not the proprietor of the *Independent* or connected with that paper in any way,' he declared. Technically this was correct: the bond connected him with Hartley, not the paper. Rival editors were not fooled. They continued to take great delight in identifying Robinson 'by the curiously vague ideas of the construction of the English language' which ran through articles in the *Independent*.

[4]

During the months immediately following the transfer of the newspaper, Robinson was not unduly concerned with politics. There were no elections in the offing; his mind was on other things. On 3 October, 1877, he was married. The wedding was, in some ways, unexpected. He was thirty-seven and had long since been written off as an unprincipled womaniser: too mean or too promiscuous to marry. Gossip had it that no female was safe with him. His favourite occupation was, it is said, to stand at the door of his office leering at passing girls. 'Although he looked as cold as a fish', claimed an old Kimberley hand, 'there was no denying his admiration for the fair sex . . . and wasn't he a Don Juan . . . he did run after the petticoats.'

Much of this talk was, of course, inevitable: the sort of tittle-tattle about any rich bachelor of long standing. Nevertheless, his marriage to an eminently suitable local girl came as something of a surprise. Less surprising perhaps was the frugal way in which the nuptials were celebrated. Never one to squander his money, Robinson apparently did not consider that his marriage warranted undue expenditure.

His bride was Elizabeth Rebecca Ferguson. Auburn-haired and pretty, she was the eldest daughter of James Ferguson, one of the pioneers of the diamond diggings. The wedding ceremony was performed in the bride's family home by the Resident Magistrate. A number of friends attended the ceremony but, apart from a conventional toast—proposed by Robinson's new friend, Dr William Murphy—there were no festivities. Nor was there anything lavish about the honeymoon, which was spent at Klipdrift (or Barkly as it was now called) the centre of the old river diggings. The only thing that now distinguished the former Klipdrift was the fact that it was 'already half deserted'. This might have made it an attractive spot for a honeymoon but, for a man as rich as Robinson, it was hardly an inspired choice.

It was left to Robinson's friends to make a bit of a splash. Three days after the wedding it was announced that: 'The many friends of the bridegroom intend celebrating the event by a banquet to take place this evening, at Mrs Jardine's Hotel.' There is no indication of who paid for this feast. Robinson was on his honeymoon.

CHAPTER SEVEN

ASPIRATIONS

BY 1877, a mere six years after the start of the dry diggings, Kimberley had become the second largest town in South Africa. Census figures for that year show the total population of the Kimberley district (including De Beers, Dutoitspan and Bulfontein) to have been 18,000: 8,000 whites and 10,000 'non-whites'. The mushroom growth of the town never ceased to amaze. Visitors, travelling across the flat and barren veld, found it difficult to believe that such a godforsaken spot could have yielded such riches and attracted so large a community. Far from producing fortunes, the parched countryside surrounding Kimberley seemed incapable of sustaining life.

'I do not think that there is a tree to be seen within five miles of the town,' declared Anthony Trollope, who arrived at the diamond fields in 1877. 'When I was there I doubt whether there was a blade of grass within twenty miles, unless what might be found on the very marge of the low water of the Vaal river. Everything was brown, as though the dusty dry uncovered earth never knew the blessing of verdure. To ascertain that the roots of grass were remaining, one had to search the ground.' Kimberley was, indeed, far removed from Barchester.

The visit of the famous novelist was a landmark in the history of the town. Men of letters were few and far between in Griqualand West. Trollope's descriptions of Kimberley are hardly flattering: he found it hot, enervating and hideous. The dust and the flies almost drove him mad. 'Dust so thick', he declared, 'that the sufferer fears to remove it lest the raising of it may aggravate the evil, the flies so numerous that one hardly dares to slaughter them by the ordinary means lest their dead bodies should become noisome.'

He found the living conditions at Kimberley equally unpleasant. The mean-looking corrugated iron houses, in which both rich and poor lived, were devoid of comfort. With all materials having to be transported to the town by ox-wagon, their could be little or no ornamentation. Lath and plaster for ceilings, for instance, was practically unheard of. 'The rooms are generally covered with canvas which can be easily carried,' he explained. 'But a canvas ceiling does not remain clean, or even rectilinear. The invisible dust settles upon it and bulges it, and the stain of dust comes through it.' Nor was the town itself a pretty place. Its most notable feature was its square and this was pointed out to Trollope with pride. He was not impressed. 'In Kimberley', he wrote, 'there are two buildings with a storey above the ground, and one of these is in the square. This is its only magnificence. There is no pavement. The roadway is all dust and holes. There is a market place in the midst which certainly is not magnificent. Around are the corrugated iron shops of the ordinary dealers in provisions. An uglier place I do not know how to imagine.'

Yet, for all his aesthetic distate, Trollope was extremely impressed by Kimberley. Not with its houses, or its streets, or with the way its citizens made their money. He considered grovelling in the earth for diamonds a degrading occupation. When he learned that women and children took a hand in diamond sorting he was disgusted. 'I thought', he exclaimed, 'that I could almost sooner have seen my own wife or my own girl with a broom at a street crossing.' What did excite him was the Kimberley Mine: the huge hole around which the town had developed—the hole which had replaced the Colesberg Koppie. He visited the mine during the day and marvelled at its 'peculiar strangeness' in the moonlight. He was lowered 230 feet to the bottom of the mine and clambered about inspecting the way in which the 408 claims—some owned by individuals, some by companies—were separated from each other. He was intrigued by the network of wires that carried the buckets of earth to the surface. 'Wires are stretched taut from the wooden boxes slanting down to the claims at the bottom—never less than four wires for each box, two for the ascending and two for the descending bucket. As one bucket runs down empty on one set of wires, another comes up full on the other set.'

All this, however, was incidental to what he considered the real wonder of Kimberley. He had little time for diamonds as such, but he rejoiced in the workings of the capitalist system. In his opinion, the flesh-pots of Kimberley were succeeding where missionaries had failed; in the glitter of the diamonds he saw reflected the 'civilisation of the Savage'. 'The simple teaching of religion has never brought large numbers of Natives to live in European habits; but I have no

doubt that European habits will bring about religion . . .', he maintained with smug Victorian hypocrisy, 'when I have looked down
into the Kimberley mine and seen three or four thousand of them at
work—although each of them would have stolen a diamond if the
occasion came—I have felt that I was looking at three or four
thousand growing Christians.' This it was that made him hail
Kimberley as 'one of the most interesting places on the face of the
earth'.

Anything was possible once the problem of 'civilising the savage'
had been solved. It could lead not only to the spread of Christianity
but, what was equally important, to the extension of British enterprise. In the diamond mines at Kimberley, Anthony Trollope caught
a glimpse of a great colonial dream.

[2]

This dream was shared by the twenty-four-year-old Cecil Rhodes,
who arrived back in Kimberley shortly after Trollope had left. Like
Trollope, Rhodes was beginning to regard Kimberley as something
more than an ugly collection of corrugated iron shacks. Although he
did not share the novelist's pious sentiments and cared little about
civilising the savage, he was fully aware of the role Kimberley could
play in extending British influence and this, to him, was as holy a
cause as the spreading of Christianity. It was something to which he
had given a great deal of thought. But it was not until he returned to
Africa at the end of 1877 that his thoughts took a more practical
turn.

For the best part of 1876 and 1877, Rhodes had been kept busy
at Oxford. His studies, like his way of life at this period, were
extremely erratic. Unable to knuckle down to disciplined reading,
he saw no point in attending lectures regularly. 'Now, Mr Butler,'
he replied to a tutor who had remonstrated with him, 'you let me
alone and I shall pull through somehow.'

Part of the trouble, of course, was his preoccupation with his
business concerns in Kimberley. He was forever writing to his
partner, Charles Rudd, discussing the pros and cons of buying new
claims, giving advice on pumping machines, speculating on the
effect of a political crisis on the diamond market, describing the
interviews he had with secretaries of rival companies and telling of
his visits to the diamond merchants of Hatton Garden. This pre-

occupation tended to set him apart from his fellow students. 'The impression he left on his contemporaries', it is said, 'was that of a shy and thoughtful man, who said little and thought much.' It is easy to understand how the moody Rhodes could have made such an impression. Never a good mixer, he was slightly older than the average undergraduate, had come, not from a public school but the Kimberley mines, and there were few at Oxford with whom he could share his business interests. All the same, it is misleading to think of him as shy. Rhodes was nervous, excitable, an introvert with a high squeaky voice, but he was not shy. Anything that Rhodes did he did deliberately. If he kept himself to himself it was because he preferred it that way. When he wanted to assert himself he was quite capable of doing so.

Rhodes made some friends at Oxford. They were mostly men who, like himself, were slightly older and more serious-minded than the general run of students. Such friends did not find him reserved or reticent. He had, when he wanted to exert it, a magnetism which drew those he wanted to attract. He could sparkle in conversation and win people round to his way of thinking. 'He certainly had the power not only of driving his ideas home, but warmly attaching to him the men who enjoyed his confidence', recalled one of his Oxford friends, 'and in my judgment he was not only a splendid Imperialist, but a most attractive personality.'

The Imperialism came later. Strangely enough, Rhodes's passionately held conviction that the English should colonise the world was not apparent to his contemporaries at Oxford. This was, perhaps, because most of them shared his belief and one Imperialist among so many could not hope to impress. Everyone at Oxford, in the 1870s, knew the creed and language of Imperialism. It had been taught to them by John Ruskin who, in his famous inaugural lecture of 1870, had urged them to make their country 'again a royal throne of kings, a sceptred isle, for all the world a source of light, a centre of peace. . . .' He had gone on to tell them how this was to be accomplished. 'This is what England must either do or perish,' he said; 'she must found colonies as fast and as far as she is able, formed of her most energetic and worthiest men; seizing every piece of fruitful waste ground she can set her feet on, and there teaching these her colonists that their chief virtue is to be fidelity to their country, and their first aim is to be to advance the power of England by land and sea. . . .'

For the young and idealistic this was heady stuff. Ruskin left an indelible impression on a generation of undergraduates. Those who came to Oxford later read the speech and were fired by it. Rhodes admitted to being, like so many others, a disciple of Ruskin's. But if he was one of many, there were not many like Rhodes. His was not merely a fashionable enthusiasm. For him Ruskin's poetic exhortation was not a matter for discussion but a call to action.

There was, in fact, little new in what Ruskin had said. Rhodes had heard it all more crudely expressed in Kimberley. South Africa, with its racial antagonisms, was then, as later, a breeding ground for aggressive Imperialism. Chauvinistic Britishers had played no less a part in the diamond fields' agitation than had those who had opposed the Imperial authorities. It was simply a matter of where one's sympathies lay. Rhodes had kept quiet during the rebellion but there can be no doubt as to which side he supported. There is reason to think that he had read Ruskin's address long before he went to Oxford: it would have been obligatory reading for exiled patriots and was probably passed around Kimberley. A little over a year after Ruskin had delivered his inaugural lecture, Rhodes suffered his first heart attack and was taken by his brother Herbert to recuperate in the Transvaal. While in the Transvaal Rhodes had thought of death and the purpose of life; this had led him to make his first will. Writing on an odd scrap of paper, the young digger had bequeathed his small fortune to the Secretary of State for Colonies: the money was to be used for the extension of the British Empire. 'This', Ruskin had said, 'is what England must either do or perish.' The poetry of Oxford gave romantic form to the blusterings of Kimberley.

But if Rhodes was waving the Union Jack before he went to Oxford, the university undoubtedly confirmed his prejudices. At Oxford it was that he thought out his peculiar philosophy of Imperialism. Not only Ruskin, but Aristotle, Marcus Aurelius and Gibbon contributed to the end result. He was influenced also, like many of his generation, by Winwood Reade's *Martyrdom of Man*: a book which preached the importance of man's own efforts unaided by supernatural powers. Life at the diamond diggings had long since demolished the teachings and taboos of the Bishop's Stortford rectory and Winwood Reade swept away what remnants of Christianity he had retained. Religion continued to interest him, but his faith was shaken: he gave God a fifty-fifty chance of existing.

Rhodes's belief in the superiority of the English race and the need for man to be self-reliant, formed the basis of his philosophy. According to W. T. Stead, it was shortly before his return to Kimberley in 1877 that he wrote a draft of 'Some of my ideas'; a curious document in which he tried to clarify his thinking on life and its ultimate purpose.

This is how his mind worked at the age of twenty-four:

'It often strikes man to enquire what is the chief good in life. To one the thought comes that it is a happy marriage, to another great wealth, to a third travel, and so on; as each seizes the idea, he more or less works for its attainment for the rest of his existence. To myself, thinking over the same question, the wish came to render myself useful to my country. . . . I contend that we are the first race in the world, and that the more of the world that we inhabit, the better it is for the human race. I contend that every acre added to our territory provides for the birth of more of the English race, who otherwise would not be brought into existence. Added to which the absorption of the greater portion of the world under our rule simply means the end of all wars.'

Crude, arrogant and clumsily expressed, this was Rhodes's credo. There were later refinements but, basically, he never deviated from the theories outlined in 'Some of my ideas'. The impact of these undergraduate musings was to be felt by millions.

To Victorian England, proud, confident and self-righteous, Rhodes's sentiments were noble; today they are recognised for what they are—a shallow rationalisation of national aggrandisement. This is not to say that Rhodes was insincere. Nineteenth-century nationalism was born of an idealism far removed from the debased aspirations of modern racialists. Rhodes truly believed that England alone could establish justice, promote liberty and ensure peace throughout the world. He saw nothing shallow in his reasoning. Misguided he undoubtedly was, but his motives were not mean. It was his sincere belief in England's manifold destiny which set him above the moneygrubbing moguls of Kimberley. Few of them could appreciate his vision. They were to admire his audacity but were bewildered by his ambitions: there was so much else that could be done with money. Nevertheless, his oddity came to be accepted. Barney Barnato summed it up. 'Some people', he told Rhodes, 'have a fancy for *this* thing, and some for *that*; *you* have a fancy for making an Empire.'

As yet that fancy was in the embryonic stage. Conceived at Oxford, it first took tangible shape in Kimberley. Rhodes went to South Africa for the Long Vacation of 1877. He arrived in Kimberley on 7 August, having spent the long dusty coach journey from Cape Town in the company of an officer of the Royal Engineers, Captain (later Sir Charles) Warren. They had got on well together. Warren, a keen Bible student, had been intrigued to find Rhodes swotting from a divinity cram-book for an Oxford examination. This had led to a friendly argument about religion: a subject on which they agreed to differ. 'He had his views and I had mine', says Warren, 'and our fellow-passengers were greatly amused at the topic of our conversation—for several hours being on this one subject.' Warren had recently been appointed as a special commissioner to Griqualand West for the purpose of investigating land claims. It did not take him long to discover that Rhodes had a reputation in Kimberley for more than his religious knowledge. 'He is', noted Warren in his diary a few weeks later, 'accredited with a long head.'

Rhodes was kept very busy during his stay in Kimberley. The activities of the Rudd-Rhodes partnership now extended far beyond their water-pumping contracts. A great deal of money could be made in Kimberley from ancillary mining enterprises, but there was no escaping the fact that it was in diamonds that the substantial fortunes were made. To own and work claims was the ambition of every Kimberley entrepeneur; Rudd and Rhodes, for all their profitable side-lines, were no exception.

Starting with a valuable block of claims at Baxter's Gully in the Old De Beers Mine (the mine which had developed, not at Coles-berg Koppie, but around the De Beers farm-house) they had gradually increased both their holdings and their influence. At one time, it is said, they had the opportunity of buying the entire mine for a mere £6,000. They had been forced to decline the offer because 'they could not afford the capital as well as the licence fees'. If this is true, they were to regret it later. At the time, however, they had not been deterred. By taking others into their partnership they had extended their mining interests and were now claimholders of considerable importance. Charles Rudd, of course, had been largely responsible for the growth of their diamond concern (he had handled the direct negotiations) but Rhodes, both by letter and on his visits to Kimberley, was actively involved in every move the partners made.

His visit in 1877 was no exception. Those at Oxford who regarded him as diffident and taciturn would undoubtedly have been surprised at the forceful way he conducted business in Kimberley. His voice was heard both at board meetings and at public meetings. Shy was the last thing anyone in Kimberley would have thought him. He was not only heard but he was listened to with respect; he seemed to be everywhere at once.

In fact, he crammed too much into too short a time. His heart could not stand the strain. Sometime during this visit he suffered another heart attack: an attack which left him shaken and frightened. He was staying, as he always stayed, in the bachelor quarters of the 'Twelve Apostles'—the young men with whom he had shared from his earliest days at the diggings—and his behaviour after his heart attack was most strange. 'His friends,' says Sir Lewis Michell, 'once found him in his room, blue with fright, his door barricaded with a chest of drawers and other furniture; he insisted that he had seen a ghost.'

Rhodes was incurably superstitious and no doubt believed what he said: but the spectre which really frightened him was the glimpse he had had of death. For death, to him, meant not only the end of his own life but the end of those high-flown ideas which he had so recently committed to paper. Nothing, perhaps, illustrates his sense of mission more effectively than his reaction to the possibility of his life coming to a sudden end. Had he been the egocentric, power-seeking monster that some of his detractors make him out to be, his thoughts would hardly have taken the turn that they invariably did at such times. He saw death not in terms of his own mortality, but as the end of his championship of a far greater cause. Not that he exaggerated his own importance and merely wished to perpetuate his name. His first will had been made on a scrap of paper in the Transvaal; he was then an obscure digger with an unspectacular fortune; he could hardly have hoped that the small bequest he made for the extension of the British Empire would ensure his immortality. No, his thoughts were not of himself but of the role which he was convinced his country must play for the peace of the world.

He had tried to formulate this belief in 'Some of my ideas'. Now, shaken by the thought that he might soon die, he attempted to express himself more purposefully. He made a second will. This time he had more money to leave and a guide to how it should be spent. His second will was more comprehensive than the first and decidedly more peculiar.

Written in the sweaty heat of a Kimberley shack, it sets out to re-order the world. He dated it 19 September 1877 (the day he was due to return to England) and described himself as Cecil John Rhodes of 'Oriel College, Oxford, but presently of Kimberley in the Province of Griqualand West, Esquire.' As executors he named Lord Carnarvon, the British Colonial Secretary and Sidney Shippard, the Attorney General of Griqualand West. These two unsuspecting gentlemen were to be responsible for the establishment of a Secret Society, whose aim and object would be the extension of British rule throughout the world. No less. A clandestine and dedicated brotherhood was to bring the entire continent of Africa under its sway and to populate South America, the Holy Land, the seaboard of China and Japan, the Malay Archipelago, the islands of Cyprus and Candia, and any islands in the Pacific not possessed by Britain, with British settlers. As if this were not enough for any underground movement to be going on with, the Society was also instructed to recover the United States for Britain, consolidate the Empire and inaugurate a system of Colonial Representation in the Imperial Parliament for the foundation of 'so great a power as to hereafter render wars impossible and promote the best interests of humanity'.

Having signed this will, Rhodes returned to Oxford.

[3]

The theatre long remained the most popular of Kimberley's respectable entertainments. Despite the incessant noise from the bar, the heat of the corrugated iron structure, and the total lack of elegance—both on and off stage—a night at the theatre was regarded as a social event. 'Here', it was said, 'was a chance for the fair ladies of Kimberley to rustle their dresses, display their jewels, diffuse their scents, show their charms, and tantalisingly tickle the amorous proclivities of the dude, the digger and the dealer.'

But the audience was not merely skittish. There were times when groups of rowdies at the bar would heckle the performers; or a row would break out in the foyer and send the audience scuttling. Even the actors could not always be trusted to remain aloof from a shindy

Cecil Rhodes as an
Oxford undergraduate

Jagger Library, University of Cape Town

The Mayor of Kimberley:
the steely-eyed J. B. Robinson
in the early 1880s

South African Library

The Dashing Young Diamond
Merchant: an early, little known
photograph of Barney Barnato
taken by J. E. Middlebrook
in Kimberley

Alexander McGregor Museum, Kimberley

The Debonair Solly Joel,
youngest nephew of
Barney Barnato

South African Library

Earliest photograph of New Rush yet traced, showing part of Main Street nearest the mine. Photographed by H. F. Gros late 1871. On the left is the notorious London Hotel and Billard Room later owned by Harry Barnato. The New Rush Mine is in the foreground

Barnato Buildings, Kimberley (Diamond Market) photographed during the 1880s. Isaac Joel was arrested in these buildings in 1884

Alexander McGregor Museum, Kimberley

Cecil Rhodes (right) aged 28 with Frank Orpen on taking their seats in the Cape House of Assembly

Jagger Library, University of Cape Town

Cecil Rhodes, standing with umbrella; J. B. Robinson, hand in pocket; Sidney Shippard, Attorney-General, seated

Jagger Library, University of Cape Town

Barney Barnato
in middle age

Electioneering in Kimberley, 1888. Barney Barnato, with
supporters in his carriage complete with liveried postillions

A typical Kimberley
scene in the 1880s,
showing teams of oxen
in the Market Square

Kimberley Public Library

A group of De Beers directors: Cecil Rhodes and
Barney Barnato seated with legs crossed, centre;
Woolf Joel extreme left

Jagger Library, University of Cape Town

The famous cheque for £5,338,650, paid by De Beers
for the Kimberley Mine in July 1889

South African Library

in the auditorium. One of the more memorable occasions when all theatrical conventions were swept aside, was at the Lanyon Theatre on 15 June 1878. On that night, a special benefit for the manager of the theatre, Mr Seymour, was performed. There were various items on the bill, the last being an excerpt from 'Othello' with Seymour himself playing Iago. Seymour was popular and the house was packed but, unfortunately, the evening's entertainment was ruined by what one newspaper described as 'a somewhat disagreeable *contretemps* between an amateur exponent of the dramatic art and a few of the unfortunate audience'.

All had gone well until the last act. Then, when Othello strode onto the stage looking, it is said, more like 'an Ethiopian minstrel on the Margate Sands than Shakespeare's heroic Moor', the audience burst out laughing. Scowling at them, Othello plunged into a long soliloquy which became more and more hysterical as the giggling and the cat calls grew louder. He came to the line 'Unhappy that I am black... ;' 'Then go and wash your face,' shouted someone in the front row. That did it. The tragic Moor was replaced by a black-faced, furious Barney Barnato who stormed to the footlights and challenged the hecklers to show themselves. Claiming that he had only agreed to play Othello to oblige Mr Seymour, he threatened to 'deal with' those who were interrupting him at the end of the play. He was as good as his word. As soon as the curtain came down, he dashed to the front of the theatre, caught the man he considered to be the ringleader of the hecklers and beat him up.

At one time the incident would not have caused much comment. It was by no means the first time a fight had broken out at the theatre; Barney Barnato was known to be quick with his fists. Nobody would have bothered much about a show of temper. But that was before the Barnato brothers had become successful claim-owners. With claims bringing in an estimated £1,800 a week, Barney was no longer regarded as a harmless young scamp whose tantrums could be overlooked. He was a man of property and his bouts of temper were no longer indulged.

To some it looked as if he was trying to throw his weight around and needed to be taken down a peg or two. 'We can fully sympathise with Mr Barnato being publicly ridiculed,' said a theatre critic, 'but no one can overlook so great a want of decency as daring to threaten individuals who have paid their fees of admission, when no agreement is made beforehand whether they are to laugh or cry. Mr

Seymour should be sufficiently experienced to know that such scenes are not likely to enhance the popularity of his establishment, which must have proved most discomforting to those ladies present, and we are much surprised he should have allowed a renewal of the offence after the fall of the curtain. As Mr Barnato explained he only appeared to oblige the management, we will deal with his rendering of the character as mildly as possible, by stating that it was simply fearful, and would remind him the audience are supposed to know nothing of the obligations existing, but visit the theatre to be either enlightened or amused, and are perfectly justified in resenting what must be termed a palpable presumption.'

This could not have pleased Barney. He appears to have been under the impression that a claque had come to the theatre for the sole purpose of making him look ridiculous. He may have been right. Like most amateur actors in Kimberley, he had rivals who would go to great lengths to spoil his performance. But the significance of the incident lies not so much in how it was inspired as in Barney's reaction to the thought that there was a conspiracy against him. The pent up fury which broke over the heads of the audience and led him to assault one of them was not merely a matter of artistic temperament. It was an early indication of a deep-rooted flaw in his character.

Barney, it is true, had been reared in a tough school. Whitechapel and his early days in Kimberley had taught him to look after himself; his father's maxim 'hit first and ask questions later' had been taken to heart. All the same, he was not aggressive by nature. He rarely went looking for trouble; there was nothing of the opinionated bully about him. In business he could be ruthless and he loved a good fight but, in his personal relationships, he was fairly easy-going. He could take a joke against himself and enjoyed nothing more than a good-natured slanging match. He seemed as imperturbable as he was impudent. Only when he thought he was being deliberately harassed, that others were ganging up on him, did this nonchalant façade crack. Then he became neurotic. The mere suggestion of organised hostility made him lose reason and act in a frenzy, as he had done at the Lanyon Theatre. That his persecutors were supported by the press only made matters worse: another dimension was added to the conspiracy. He could stand up to the toughest individuals but he could not bear collective ridicule. Time did nothing to harden him. As his career progressed, so did public criticism and so, unfortunately, did his paranoia. His outburst as Othello was symptomatic of a basic insecurity.

[4]

The publicity given to Barney's tantrum could hardly have pleased his brother. As an established claim-owner, Harry was doing his best to live down his raffish past. The days of Signor Barnato, the Great Wizard, were decidedly over. In May 1878 he had sold the London Hotel and with it discarded many of his old cronies. He was also trying to drop the name Barnato. As a stage name, and even as a business name, Barnato had served well enough but it was somewhat lacking in dignity. In private life he was now plain Henry Isaacs, committee member of the Griqualand West Jewish Congregation, active supporter of the Jewish Benevolent Society, and a generous donator to charities of all kinds. For the most part he kept himself out of the public eye and concentrated on building up the Barnato Brothers' mining concern and diamond buying business. He was aiming, it seems, to be accepted by Kimberley's respectable Jewish society. He got very little help from his brother.

Barney was an emotional rather than a devout Jew. 'He believed in the Jewish religion,' says Lou Cohen, 'but never gave it a thought; he respected the synagogue, though seldom entered one.' Every bit as generous as his brother, Barney tended to be less discriminating in the charities he supported. For instance, as B. B. Isaacs he would give a large donation to a fund for aiding the persecuted Jews of Poland and shortly afterwards, as Barney Barnato, he would promote a special theatrical performance in support of the 'Building Fund of the Catholic Chapel'. Try as he might, he could never imitate Harry's new-found aloofness. For one thing, he was far too jealous of his popularity among Kimberley's *hoi polloi*; for another, his private life was far too unconventional.

Exactly when Barney set up house with Fanny Bees is not known. It was probably in the mid-seventies. Fanny was the daughter of John Bees, a tailor who had worked in Simonstown, the naval base at the Cape, until shortly after the rush to the dry-diggings when, like many another hopeful tradesman, he had trekked to the diamond fields. His career had been short lived and unlucky. For the first few years the family (there were eight children) had eked out an existence at Bulfontein, living in a tent and cooking over a camp fire.

As soon as she was old enough Fanny took a job to help the family finances. She was still in her teens when she went to work in Kimberley as a barmaid. In her spare time she acted in minor roles at the Theatre Royal; it was probably here that Barney first met her.

She was a very attractive girl. Taller than Barney, with a generous mouth, flaring nostrils, a honey-coloured complexion and dark springy hair, Fanny Bees had an aura of voluptuousness. It is easy to understand how Barney was drawn to her. More surprising is her uncritical response to the urchin-faced Barney. Not even Barney's best friend would have called him good-looking. P. Tennyson Cole, the portrait painter, was to say that Barney Barnato was 'the plainest man I have ever known'. Fanny did not agree. 'Perhaps I am biased because I loved him,' she said, 'but strange as it may seem [he] to me seemed a very handsome man.'

They undoubtedly loved each other. In a freer atmosphere they might have settled down immediately to an uncomplicated married life. But life in Kimberley was restricted; gossip and prejudice were rife. It was not merely that Fanny and Barney were known to be of different religions; rumour had it that Fanny, who claimed to be of Huguenot descent, was of mixed blood. She was said to be 'an Afrikander of the St Helena type' and this, in the terminology of the day, meant that she was a Cape Coloured. If this is true (and photographs of Fanny make the rumour, at least, understandable) it might well explain why they lived together many years before they were married.

Years later Barney was to claim that they were secretly married at a civil ceremony in 1877 but, as Fanny was then a Christian, they had been unable to make this known as he did not wish to offend his parents. This seems very doubtful. There is no record of this marriage and, if it had taken place, it is surprising that when they were eventually married in a London registry office in 1892, the ceremony was again a civil one and not the religious one that had been neglected (Fanny had by then been converted to the Jewish faith). At the time of the London marriage Barney was described as a bachelor and his bride as 'Fanny Bees, Spinster'.

The difference in religion between Fanny and Barney would not have been a serious obstacle. Marriages between Jews and Gentiles were by no means uncommon in Kimberley. Barney was far from punctilious in his religious observances and he must have known that Fanny would have been more readily accepted by the majority of the

Jewish community as his wife than she was as his mistress. However, if Fanny were coloured, things would have been more difficult. There was no legal bar to a racially mixed marriage in South Africa at that time—not until 1902 were the first miscegenation laws enacted—but the social stigma from such a marriage could not easily be ignored. Many men in Kimberley had coloured mistresses, but only those willing to become social outcasts had coloured wives. If Fanny was not coloured, she certainly looked coloured and this would have been sufficient to ostracise her and Barney as man and wife.

Barney, for all his lack of convention, was not yet ready to defy society to such an extent. All the same, his loyalty to the woman he loved is greatly to his credit. Not only did he make no secret of his liaison with Fanny but, as soon as he was rich and powerful enough, he married her and forced the bigots to accept the marriage. Not every digger in Kimberley showed such fidelity.

Fanny always kept herself very much in the background. Of the hundreds of anecdotes that are told about Barney Barnato, few acknowledge his wife. For all that, Fanny was not without ambition. She wanted Barney to make a name for himself, not only as a moneyed man, but as a public figure. It is said, for instance, that she was responsible for Barney taking the unlikely step of standing for the Kimberley Town Council in 1878. Certainly his candidature came as a shock. One newspaper went so far as to say that he had only been nominated 'for the mere sake of bringing the whole Municipal Council into contempt'. Not for one moment did anyone think that Barney was serious. But he was. And he soon let them know it. A couple of days later the newspaper apologised for saying that his candidature had been a joke.

[5]

Barney's first venture into public affairs is instructive. It provides a glimpse of the growing Barnato power. The firm of Barnato Brothers was still in its infancy, but it was sufficiently well established to swing a local election.

Admittedly all that was necessary to do this was money. The Kimberley elections were notoriously corrupt. 'It is well known', it was said of the election Barney fought, 'that . . . some of the contests were largely affected by voters who came to the polling stations provided with receipts for rates which had been openly and without concealment paid by the aspirant for Municipal honours or his accredited agents. Now there can be no doubt that this is bribery in its most direct and objectionable form.' There can also be no doubt that Barney was one of the offenders. He not only headed the poll in Ward 5 but saw to it that his running mate—a man named Lawrence —was also elected. They were opposed by George Bottomley, a somewhat pompous teetotaller who, with the help of the Good Templars' association had recently been elected to the Legislative Council. Mr Bottomley was now determined to air his abstemious views in the Town Council. Barney had other ideas.

Voters were slow in coming to the polling booth in Ward 5. Barney's colours, blue and white, were very much in evidence on the morning of the election, but there was no rush to vote for 'the youthful candidate'. The first votes all went to Mr Bottomley. 'After a little while, however,' it was reported, 'a file of ten men appeared, and in answer to the usual question replied "Plump for Barney". Votes poured in and Mr Barnato soon had a commanding lead. By 12 o'clock seeing his position assured, he set to work on Mr Lawrence's behalf, polled the latter from third to second place, and maintained the relative positions until the close.' The ginswiggers of Kimberley were greatly relieved at Bottomley's defeat. Barney's success was attributed euphemistically to 'the excellent organisation of his working committee'. Not until later were more accurate descriptions found for the rigged election.

For all that, Barney proved a conscientious councillor. He dignified himself by following Harry's example and reverting to the name of Isaacs (although he was often reported as Barnato); he rarely missed a council meeting.

In his election manifesto he had promised to give urgent attention to the infamous drainage and water supply of Kimberley and at an early meeting he was, in fact, elected to the sanitary committee. But his heart was not really in water affairs. The workings of the Municipal market were far more to his taste. His experience at Petticoat Lane convinced him that not only could he make the market pay but that he could set an example to the council's employees. Many a raised eyebrow was caused by his antics in the market place.

'An edifying spectacle may be observed on the Market every morning,' it was reported in April 1879. 'A full fledged Town

Councillor, clad in check trousers etc., touting for the assistant Marketmaster, holding up cabbages and other vegetables to the gaze of an admiring public, and finally returning home with a cabbage under each arm, a pocket full of carrots, or some other presents contributed by grateful sellers, who seem to appreciate the efforts of this worthy controller of our roads and morals to get them the highest prices for their products.' Bribery, it seems, worked both ways.

Barney was in his element on the Town Council. He adored an audience, be it in the theatre, the market place or the council chamber. If he had a fault as a councillor, it was his tendency to let his tongue run away with him. His speeches were often as irrelevant as they were irreverent. He regretted that council meetings often clashed with his work at the mine and that consequently he sometimes had to cut his speeches short. In fact, one of the first resolutions he moved was that council meetings be held in the evening instead of in the heat of the afternoon. The Mayor was quick to oppose this. Looking fixedly at Barney, he pointed out that 'members who had had a glass of wine for dinner, would want to air their eloquence before the public'. A poker-faced Barney then 'gravely assured the Mayor that he had no wish to air his eloquence'. The motion was, nevertheless, defeated.

For all his fooling, Barney did a competent job. At the elections held at the end of 1879 he was returned unopposed. He did, however, play a part in another municipal contest; one which first brought him into contact with J. B. Robinson and which, indirectly, involved Cecil Rhodes. It marked the humble beginnings of big events.

CHAPTER EIGHT

KIMBERLEY'S MAYOR

AT the beginning of 1880, an American actor, Stephen Massett, visited Kimberley. He was touring South Africa with a theatrical company and supplying an American newspaper with descriptions of his travels. Kimberley taxed his literary powers. 'I have never', he declared, 'visited a place so difficult to describe.'

His first impression was that the town, with its one-storeyed iron houses and straggling streets 'looked like San Francisco in the early days, or some of the interior California mining towns'. But this impression soon faded. For all its primitive appearance, Kimberley lacked the rough and tumble of the Far West. Massett was impressed by the facilities: 'Here,' he said, 'are three or four large, and even elegant haberdashers or dry goods stores, stationer's shops, chemists and druggists, boot and shoe makers, toy shops, a well appointed club, called the "Craven", billiard saloons, all lighted up brilliantly.'

But perhaps more than anything, what pleased him was the care-free spirit, the lack of violence and fear, the feeling of well-being that seemed to pervade the town. This he thought truly remarkable. 'Two things seem to possess everyone—viz., diamonds and money-getting, and the idea of clearing out as soon as your "pile" is made', he wrote. 'But in spite of the discomforts of living, of the privations you are subjected to, everyone seems jolly, happy, and lighthearted. The people are kind, hospitable, generous and good-natured, and the stranger welcomed and treated with the utmost consideration. Money is plentiful and they spend it freely; there is nothing mean or small about the Kimberleyites . . . whatever the difficulties—and they are many—that are encountered in getting to this—once upon a time—barren wilderness, they are quickly forgotten in the kindly and cordial welcome you receive from its free-hearted, liberal, enterprising and generous people.'

Much of this expansiveness was derived from Kimberley's orderly way of life. The town was well run, its civic institutions were efficiently organised (he thought the prison the most perfect of its kind in South Africa) and law and order prevailed. He gave his

American readers a break-down of Griqualand West's system of government, ending with the administration of Kimberley itself. 'There is a Municipality for the town of Kimberley,' he explained, 'administered by a Town Council, who elect their Mayor: J. B. Robinson is the present popular chief.'

Massett's pen-picture of Kimberley reflects the town as it appeared to a chance visitor: orderly, happy-go-lucky, and friendly. Like all such fleeting impressions it is both illuminating and deceptive. To the outsider Kimberley might indeed have appeared a harmonious, if primitive, outpost of civilisation; to its inhabitants it was, like most provincial towns, a hotbed of petty intrigue, back-biting and rivalries. Commonplace as such parochial faction-fighting was, in Kimberley the diamond mines gave a distinctive edge to it. Huge fortunes were being made there and even the most trivial jostling for position had a financial significance. As the mines went deeper so the running of the diamond industry became more complex; the animosities of local politics reflected a good deal more than the usual parish rivalries. Sooner or later a more powerful struggle would take place. The men of Kimberley knew this and, in their various ways, were beginning to muster their forces. There was far more to Griqualand West's politics than was reflected in Stephen Massett's happy account.

[2]

Nowhere was this jockeying for potentially important positions more apparent than in the recently elected Town Council. Mr Massett's description of the new Mayor was more conventional than accurate. Not only was J. B. Robinson's popularity open to question, but many doubted whether he was entitled to hold office at all. Civic dignity had, in fact, come to J. B. Robinson as so much else came to him: cloaked in controversy and delivered in a court of law.

Marriage had not tamed him. There had been a short period when it had looked as if a settled domestic background might change the habits of a lifetime. Some months after his hasty honeymoon at the river diggings, he had surprised everyone by taking his wife to Europe. As far as is known, this trip had little to do with his business concerns in London; it had been given over to sight-seeing excursions. This, for the penny-pinching Robinson, had seemed a transformation indeed. He had returned to Kimberley in February 1879, looking hearty, relaxed, and at peace with the world.

Unfortunately, the old surroundings soon got the better of the new man. He quickly reverted to type. The looks which had earned him the nickname of 'The Buccaneer'—the steely glance and the set mouth—had been enhanced by a great black beard, and were as unsympathetic as they had ever been; he still wore that hard, uncompromising-looking pith helmet. Nor was he any less austere or miserly. He never entered a bar other than to follow up some business deal and was known to peep through the doors first lest he be forced into buying a drink for some unprofitable acquaintance. And of course, he remained as touchy, as belligerent and as quick-tempered as ever. It was not long before he was again the centre of a scandalous public brawl.

In fairness, the fault was not entirely his. No sooner was he back than his opponents started sniping at him. Stories were spread about his marriage, cryptic paragraphs appeared in the newspapers and both his personal and his political life came under fire. Among the worst offenders were the new editors of the *Diamond News*—two brothers, George and Henry Vickers—who regarded Robinson as a rival newspaperman and fair game for attack.

For a while Robinson showed admirable restraint. He could not keep up this pose for long. Breaking point was reached when an obscure paragraph appeared in the gossip column of the *Diamond News* on 22 July 1879. No names were mentioned; the columnist merely hinted that a writer on Robinson's paper, the *Independent*, was ignoring a child who cried vainly for help. The meaning of this innuendo was never made clear. But there can be no doubt that the callous writer referred to was Robinson himself. In the next issue of the *Independent* the editors of the *Diamond News* were accused of personal invective against Mr J. B. Robinson. Nor did the matter end there.

On Monday afternoon, 28 July, George Vickers was drinking in the Queen's Hotel when Robinson stormed up to him and demanded to know if he was one of the proprietors of the *Diamond News*. 'I am,' said Vickers. 'Then you are a blackguard,' shouted Robinson. In Kimberley there was only one answer to that kind of talk. Vickers gave it. Lunging at Robinson, he gave him a punch which, it was colourfully reported, 'tapped Robinson's claret and closed up his eye and he was carried off by his friends'.

But Robinson did not give up. That same evening he returned to the Queen's Hotel determined to tackle the second brother, Henry Vickers. This time he was careful not to be caught off his guard. His lieutenant, Dr William Murphy, and some friends accompanied him and he sent Murphy over to Henry Vickers with his challenge to a fight. Vickers, like his brother, was only too willing. He asked a friend to act as his second and started peeling off his coat. However, his friend suggested that the fight should take place in some quieter place. 'This', it was reported, 'did not meet the views of Mr Robinson and his friends and while some haggling was going on Mr Robinson slipped through the crowd and struck Mr Vickers unawares on the bridge of his nose cutting Mr Vickers nose with his diamond ring and blackening his eye. . . . After the blow there was a scene between Vickers and Robinson when the police came up.'

Investigation proved the brawl to have been the result of an unfortunate mistake. Neither of the Vickers brothers was responsible for the offending paragraph, and had been unaware of it until they were attacked. The following day they went to Robinson and apologised. It was then decided that a discreet retraction should be published in the *Diamond News* and the matter forgotten.

And forgotten it might have been had Robinson not been such a braggart. He could not resist the temptation to crow. Mistaking the Vickers' apology for admittance of defeat, he published his triumph in the *Independent*. 'Stockdale Street was the scene of some excitement last evening,' ran the report. 'Mr J. B. Robinson who has lately been most unwarrantably and maliciously slandered in the *Diamond News* administered severe but at the same time well deserved punishment on the two Messrs Vickers nominal proprietors of that paper.' He should have known better. Kimberley had a third newspaper, the *Diamond Fields Advertiser*. The editor of that paper was Robinson's old enemy R. W. Murray. Murray had never forgotten Robinson's threat to horsewhip him. Now he could get his own back. He quickly got hold of an eyewitness and published a blow by blow account of the fight which his rivals were trying to play down. His report was taken up by newspapers throughout South Africa.

The unfortunate business undoubtedly upset Robinson. Once again the image he was trying to create had been badly dented. What is more, the bad publicity had come at a time when he could least afford it. For he was then on the point of making a new bid for public office: he had made up his mind to become the mayor of Kimberley. Modest as this ambition seems, Robinson regarded it merely as a stepping stone to further power. As Kimberley's leading

citizen he would have considerable influence; it would provide him
with the prestige he needed to dominate the diamond industry.
There can be little doubt that this was his ultimate aim. Brawling in a
public bar was hardly the best way to present himself for election.

[3]

In most other respects, Robinson stood a good chance of achieving
his ambition. The fact that he was not a member of the Town
Council presented no problems. He had both the money and the
means to get a seat. As a mine owner and a diamond buyer few
could equal him in the business world and his position as the power
behind the *Independent* gave him an added advantage. The *Independent* was now the most important newspaper in Kimberley. Previously it had, like its rivals, appeared three times a week; however,
towards the end of 1879 it changed its name to the *Daily Independent*
and became the first Kimberley paper to appear regularly every
weekday. Robinson's appointee, Casper Hartley, had raised both the
standard and the profits of the paper. To advertise in the *Independent*
cost as much as to advertise in the London *Times*. The value of such
advertisements was fully appreciated. 'I should imagine the income
of the *Independent*, looking at the advertisements, to be something
wonderful,' said a visitor. 'I am told that Rothschild, a celebrated
auctioneer here, pays something like £500 a month for his advertisements.' With such influential backing Robinson could count on
swaying public opinion in his favour. More difficult was the problem
of getting his fellow councillors to elect him as mayor.

Much depended on the outcome of the council elections. Several
seats were unopposed and Robinson was among those returned
without a fight. So was Barney Barnato and a number of other
financiers who could be depended upon to vote for Robinson as
mayor. Only two wards, in fact, were contested; of these the voting
in Ward 4 was expected to produce the most interesting results.
Here three newcomers had been nominated for the two seats. Two
of the candidates—Henry Chapman and William Thompson—
were known to be Robinson supporters. The third was Cecil Rhodes,
whose sympathies were extremely uncertain.

Rhodes had been late in entering the lists. Since his most recent
return from Oxford, at the end of 1878, he had been kept busy with
his mining concerns. His protracted University studies—interrupted

by illness and the constant journeying between two continents—
were almost complete. He had, in fact, to attend Oxford for only one
more term before taking his degree. Now his mind was turning to
the serious work he had set himself. Like Robinson and Barnato, he
had recognised the importance of local politics and allowed himself
to be nominated for Ward 4. Then, quite unexpectedly, he changed
his mind. Three days before the election he wrote to the Town
Clerk and withdrew his candidature.

Rhodes's withdrawal had farcical, and far-reaching, results. It
confused the entire issue of the election and landed J. B. Robinson—
inevitably—in court. In the first place his withdrawal lulled his two
Ward 4 opponents into such a state of complacency that they were
late in handing in their requisitions. The Ward 4 election was
consequently postponed for two days. Thus, with the other elections
having taken place as planned, the two Ward 4 members were
missing when the new council met to elect a mayor. But not for long.
Their postponed election having taken place on the very day of the
mayoral election, they came haring across to the Council Chamber
just in time to vote for a mayor. This, the Town Clerk assured them,
they would not be allowed to do: their election to Ward 4 had not
yet been gazetted.

The Town Clerk's ruling accepted, the remaining councillors
applied themselves to the all important business of electing a mayor.
Barney Barnato proposed Robinson. Barney had an eye to business.
The diamond industry was by no means in the majority on the
council and he recognised the importance of supporting an influential diamond merchant like Robinson. His proposal was seconded
by his old rival, George Bottomley, who approved not so much
of Robinson's business connections as of his abstemious drinking
habits. There was only one other proposal. This was for a former
mayor, John Birbeck, who was known as an old political opponent of
Robinson's. A hat was passed round for the votes. When a count was
made it was found that the two members for Ward 4 had defied the
Town Clerk and voted for Robinson. The voting slips were promptly
torn up and the hat passed round again. This time the chairman
kept a close watch on the voting and refused to accept the slips
presented by the recalcitrant members for Ward 4. The vote this
time was five votes for Birbeck and four for Robinson.

But J. B. Robinson had no intention of accepting the result. He
maintained that he had been elected fairly on the first ballot. 'Whether

Mr Robinson was elected by a combination of Good Templars and the diamond trade or not,' declared the *Independent*, 'he was elected.' Supported by Barney Barnato and those who had voted for him, Robinson applied to the High Court for the election of the new mayor to be set aside. The application was heard before Mr Recorder de Wet on 18 December.

Judgement was given two days later. As always in legal actions brought by J. B. Robinson, the case had attracted country-wide attention and the court was packed. The verdict was in Robinson's favour. The judge said: 'Mr Birbeck was a usurper and Mr Robinson should have been declared the Mayor for the ensuing year, and therefore under the circumstances I declare the election of Mr Birbeck null and void, and that Mr Robinson should be declared Mayor for the ensuing year.' The decision was not well received. But there was nothing anyone could do about it. Like it or not, J. B. Robinson had achieved his ambition and was now the Mayor of Kimberley. 'It is to be hoped', said the *Diamond News*, 'that the battle which has been fought in the High Court will not be renewed in the Council Chamber.'

Surprisingly enough it was not. J. B. Robinson was extremely proud of his position and remained so throughout his life. He was, in time, to become one of the richest men in the world and have much to boast about, but he never failed to rate his term as Mayor of Kimberley among his great achievements. In welding the council together, he displayed uncharacteristic tact and diplomacy. Staunchly supported by Barney Barnato and the other diamond merchants he managed eventually to win over his more suspicious opponents. He was so successful that even his press rivals were forced to acknowledge his newly revealed talents. 'Those who opposed Mr Robinson's election to the civic chair', declared the *Diamond News*, at the end of February 1880, 'unite now in praise of the manner in which he carries out the important duties of his office.'

Of course this pleasant state of affairs could not last. Robinson retained his hold on the council but, inevitably, fell foul of the press. Old grievances were raked up by his opponents and used to embarrass him. He was never allowed to forget the unfortunate manner in which he was elected: 'Mr J. B. Robinson who ... in consequence of a legal quibble fills the office of Mayor of Kimberley' is how he was reported. By the end of the year the long-standing feud between Robinson and the *Diamond News* was causing uproar at council meetings. It culminated in two extraordinary scenes. The first was when Robinson, shouting with rage, ordered a reporter from the *Diamond News* out of the Council Chamber. Unseemly as such behaviour was on the part of the mayor, it was nevertheless typical of Robinson.

The second scene was less characteristic. Most of the councillors had considered the expulsion of the reporter justified. A few days later they tabled a resolution expressing their support for the mayor. When Robinson rose to reply he was unusually nervous. Half-way through his speech of thanks he became so choked with emotion that he broke down and had to rush from the chamber to recover himself. Eventually he returned and stammered his way to the end of his speech. There was something very unstable about J. B. Robinson.

CHAPTER NINE

NEW ENTERPRISES

THE year 1880 was to be remembered in Kimberley for more than Robinson's controversial mayoralty. It was the year that saw the beginning of the end of the small digger.

The fusion of mining interests was a gradual process. In the early days of the dry diggings, when no-one was allowed more than two claims, diggers had worked with a few African labourers in claims separated from their neighbours by a narrow road. The road, regarded as common ground, was essential for the transporting of soil to the sorting tents at the circumference of the mine. This system had worked well enough as long as the diggings were relatively shallow. And at one time it was thought that diamonds would peter out before the mine had become too deep. One of the causes of the economic depression of the mid-seventies had been the loss of confidence when it was found that the arid yellow ground, in which the early diamonds had been unearthed, gave way to the so-called 'blue ground' which was firmer, difficult to work, and at first appeared unproductive. For many diggers, this blue ground scare proved the last straw. Life was never easy in the 'hellholes' of Kimberley and when it appeared that the ground was nearly exhausted, they gave up in despair. Claims were sold for a song and the diggings were left to those who could afford to take chances.

The blue ground, of course, proved far from unproductive. The deeper it went, the richer it seemed. But mining was more expensive. The days of the simple pick and shovel operations soon came to an end. Expensive machinery was needed: machinery which often had to be obtained from Europe and cost the earth to transport to isolated Kimberley. The inability of some diggers to meet the new demands complicated matters more. The depth to which claims were dug varied according to the claimholder's resources; this made mining not only difficult but dangerous. Those claims which went deepest suffered from the debris of those above them; the walls dividing the claims (which had once acted as roadways) became more and more precarious. Tons of reef collapsed, ruining many of the poorer diggers. Matters were more complicated by the fact that some claims were subdivided—into halves, quarters and even eighths—and digging in each claim varied accordingly.

Water seeped into the claims, which were quickly flooded, and pumping was an expensive and perpetual necessity. Few could avoid the costly legal proceedings which resulted from claimholders encroaching on a neighbour's ill-defined boundaries or purposely burrowing under a rival's territory. And the courts were continually fining diggers who neglected to safeguard their neighbour's claims against falling reef. Mining, even when it involved valuable claims, was becoming impossible for the man who lacked financial or cooperative backing.

The need to combine was obvious. The two-claim restriction was soon scrapped. Diggers were at first allowed to own up to ten claims and then, in 1876, the ten-claim restriction was abolished. This led to greater enterprise. Attempts had earlier been made at a form of amalgamation, but they had been frustrated by the limitation placed on claim ownership. Once restrictions were lifted a group of diggers, backed by a London syndicate, made another bid to combine. This bid failed because the group was unable to buy up certain claims essential to their scheme. The obstructing claims were owned by J. B. Robinson and he adamantly refused to sell out to the group. As a result, a legend arose that J. B. Robinson was opposed to amalgamation. Nothing could have been further from the truth. Robinson was wholeheartedly in favour of amalgamation; but it had to be an amalgamation which he controlled. He had, in fact, already launched his own amalgamation scheme and was to accuse his rivals of obstructing *his* company.

In December 1876, a month after the lifting of the claim restriction, Robinson published a notice inviting 'such claimholders as might be inclined to join him in amalgamating their ground for the purpose of working it upon co-operative principles'. This seemingly innocent move had brought him under widespread attack. It was generally believed that, with the help of a few other 'large capitalists', he was embarking on a financial campaign which would enable him to take over the entire Kimberley Mine. If this was in fact his intention (some doubts were expressed to as the seriousness of his proposal) it came to nothing. Several claimholders did show interest in his scheme but he was frustrated by the manoeuvrings of the London-backed group. 'Mr Robinson found', it was later announced, 'that the ground put into the company could not be connected, owing to some claims intervening which belonged to large block-holders or had been put into the [rival] company. . . . As it was essential, in order to establish a company upon the basis proposed by Mr Robinson, that the ground should be in one block, he abandoned the scheme.'

But he did not abandon it for long. Nor were others slow to follow his example.

When Anthony Trollope visited the diggings the following year, one of the most prosperous concerns in the Kimberley Mine was the firm of Messrs Baring-Gould, Atkins and Co. The head of this company, Francis Baring-Gould, had been long established on the diamond fields and had been quick to recognise the advantage of cooperative mining. By combining his claims with those of a neighbouring digger, he had not only taken control of fifteen claims but, as Trollope noted, had 'gone to the expense of sinking a perpendicular shaft with a tunnel below from the shaft to the mine—so as to avoid the use of the aerial tramway'. The shaft, in fact, did more than that: it helped to safeguard Baring-Gould's claims from falling debris. In time the sinking of shafts was to play an important part in the diamond mining industry.

The abolition of the ten claim restriction not only encouraged enterprise, it also opened the way to the international money market. Banks were now more ready to give credit and the opportunity was provided for attracting foreign investors and speculators. Groups of diggers were now able to turn over their claims to joint stock companies.

At first European investors showed little inclination to rise to the bait of South Africa's diamonds. Shareholders in the early registered companies were largely Kimberley residents. By 1880, however, at least one recently formed company had foreign backing. This was an amalgamation popularly known as the French Company. The firm had originally been founded by an itinerant trader, Sammy Marks, and his brother-in-law, Isaac Lewis. The rise of the firm of Lewis and Marks was one of the success stories of Kimberley. The actor, Stephen Massett, had outlined it for his American readers. 'Many stories of fortunes made here and lost have been told me,' wrote Massett; 'of the former perhaps "Lewis and Marks" are the best living examples. Commencing in a very small way in the "clothing" and "watch" line, they watched their chance, and by successful diamond dealing, and advancing money to claimholders, they became themselves, under the name of the "Kimberley Mining Company" with a few sleeping partners (who were wide awake, I guess!) the largest holders of the mine; at the present time holding, I am told, £200,000 of property. Upon this scene in 1875 appeared a Mr Porges, of Paris and London, coming at a time of serious crisis. Mr Porges, representing large diamond firms in London and Paris acquired for himself and friends a few valuable lots of ground, paying therefore £120,000. Then making the acquaintance of Messrs Lewis and Marks, with the assistance of a Mr Wernher, who is Porges's representative here, they amalgamated, and have formed a large French Company, called "The French Diamond Mining Company at the Cape", and this I expect will soon be put upon the London market and shares quoted.'

Mr Massett made it all seem cosy and simple. His description of the foundation of the French Company hardly reflects the ruthless methods upon which mergers were based. It was a cut-throat business. So much so, that in little over a year after Massett's visit Lewis and Marks had been 'squeezed out' of the French Company and forced to start a new trading enterprise in the Transvaal. Jules Porges & Co., based in Paris, held the bulk of the shares in the French Company and their Kimberley representative, Julius Wernher—a burly, handsome German—together with a far from sleeping partner, was soon preparing for further financial conquests.

The struggle for control of the Kimberley Mine was beginning. Two of the main contenders—Baring-Gould and the French Company—were already well established; others were just limbering up. Everyone recognised the inevitability of a single company controlling the mine. When, in October 1881, the *Independent* was accused of secretly trying to bring about an amalgamation, it replied: 'To secure the amalgamation of the Kimberley Mine will require no conspiracy on our part; come it will in the end, whether we fight against it or no.' It was merely a question of which faction would come out on top. At this stage of the game, there was certainly no shortage of contestants.

[2]

Following the amalgamation of Lewis and Marks in 1880, several
other important companies were started. What Baring-Gould and
the French Company were attempting in the Kimberley Mine,
Cecil Rhodes was determined to achieve at the De Beers Mine. His
plans were already well advanced. The Rudd-Rhodes partnership
had established a firm foothold in De Beers. By combining with
other claimholders they had built up a syndicate which controlled an
important section of the mine. On 1 April 1880, they were able to
announce the formation of the De Beers Mining Company, with an
authorised capital of £200,000. This was not, as his biographers
seem to think, Rhodes's only mining interest at this time. By the
end of 1880, Rhodes was a director of the Lilienstein Mining
Company, which operated in the Bulfontein Mine; in January 1881,
he became a director of the Kimberley Tramways Company; his
name appeared as a director of the International Diamond Mining
Company and both he and Rudd were directors of the Kimberley
Coal Mining Company.

Widespread as were his financial interests, Rhodes was content to
work behind the scenes. He was still a young man—he turned
twenty-eight in 1881—and he was prepared to allow older men to
act as figure heads for his companies. The first chairman of the
De Beers Mining Company, for instance, was Robert Dundas
Graham—a man who had been associated with Rudd and Rhodes
since 1874. Graham, a solictor by profession, was, like most men in
Kimberley, a man of many avocations. His activities were by no
means confined to the De Beers Mine. At the time that the De Beers
Mining Company was formed, he was acting as the legal adviser to
another recently established company in the Kimberley Mine. This
was the Standard Diamond Mining Company, founded with a
capital of £225,000 in £100 shares, and whose chairman was the
Mayor of Kimberley, Mr J. B. Robinson. The Standard was only
the first of several companies headed by Robinson in 1880. By the
end of the year he was known to be the chairman of the Rose-Innes
Diamond Mining Company, the Crystal Diamond Mining Company
and the Griqualand West Diamond Mining Company and was
suspected of being behind a few lesser concerns. Offices for these
newly formed companies sprang up on either side of Robinson's
diamond buying business—stretching from 61 to 65 Main-street—

and left nobody in doubt as to the mayor's financial interest in the growth of Kimberley.

The Barnato brothers were not far behind in the company launching race. Their approach, however, was somewhat more devious. On 29 May, 1880, the *Independent* announced: 'Another Amalgamation—We understand that Messrs Alfred Abrahams, B. Barnato Isaacs and Henry Cohen, of Kimberley and Dutoitspan, have been appointed buyers for, and local representatives of a new Company of the leading Continental Diamond merchants, and are prepared to buy at the highest market rates. Mr Barnato Isaacs leaves shortly to undertake the London agency.' A similar report appeared in the other newspapers. This news item was undoubtedly given out as a blind.

Barney Barnato had no intention of acting as anyone's agent; he and Harry were about to launch an enterprise of their own but, for the time being, they needed to keep their activities dark. Barney went through all the motions of leaving Kimberley. In June he put his house and furniture—including 'his brass bedstead, the best in Kimberley, and his piano, well tuned'—up for public auction. He could not, however, bring himself to resign from the council. When the time came for him to hand in his resignation, he explained that he might be in Europe only for four months. This, as he probably anticipated, brought a storm of protest from other councillors who insisted that his resignation be declined and that he be given leave. The Mayor agreed.

Barney and Harry left Kimberley in the middle of June 1880; they were away six months. (In October, Barney wrote to J. B. Robinson requesting an extension of leave.) The main purpose of this European visit appears to have been to open a London office of their firm. The office, which was to handle the marketing of their diamonds in Europe, was established at 106 Hatton Garden as 'Barnato Brothers, Diamond Dealers and Financiers'. Barney, it is said, was sadly disappointed with this, his first, visit home as a rich man. If, when he sold his house and furniture in Kimberley, he had been considering taking over the London office, he quickly abandoned the idea. Whitechapel was crowded with Russian refugees and the widespread poverty of the district was depressing. Several charities benefited from Barnato donations, but nothing came of Barney's vague plans for buying a house in a fashionable area. When Harry arrived back in Kimberley on 2 December 1880, Barney was with him.

Three months later they were in a position to embark upon their new venture. 'Messrs Barnato Brothers have started four Mining Companies within the last fortnight,' reported the *Independent* on 10 March 1881. 'The total capital subscribed was considerably over half-a-million, and covered the amount asked for many times over. The shares for all four companies are at a premium, and the undertakings may be regarded as a wonderful success.'

Success was not the word everyone used.

[3]

The long-standing dispute over Griqualand West's political status finally came to an end when, on 15 October 1880, the territory was formally incorporated with the Cape Colony. For nine years Griqualand West had been governed as a British Crown Colony. Now the territory was entitled to send members to the Cape parliament. The Cape Act which proclaimed the new Province of Griqualand West had created two electoral divisions: one for Kimberley and one for the Barkly West district. Each of these divisions was entitled to elect two members. The first parliamentary elections were announced for March 1881.

There was no immediate rush of candidates for the four seats. Election to the Cape Legislative Assembly lacked the attraction of local politics. Not only was the election contest costly but the successful candidates would be forced to spend lengthy periods in Cape Town and their business interests might suffer. To the money-conscious inhabitants of Griqualand West this was a serious consideration. While the importance of the diamond industry being strongly represented at the Cape was generally recognised, most local politicians were prepared to hand over this responsibility to others.

From the outset it was acknowledged that, whatever the competition, the Kimberley division was more likely to attract candidates than was Barkly West. Kimberley was the heart of the diamond industry, big business was centred there, and electioneering in the town was easier than canvassing the widespread rural area. As it happened, Barkly West was not contested. Only two candidates presented themselves. One was Frank Orpen, an Irishman whose family had long been active in South African politics; the other was Cecil John Rhodes.

That Rhodes should have stood for one of the new seats is not surprising. The Cape parliament was a far better sounding chamber for his grandiose ideas than was the Kimberley Town Council. He had no hesitation in accepting nomination for Barkly West. But why he chose this constituency is not known. His interests in Griqualand West were solely in the diamond industry and, while Barkly West included the river diggings, he had never been concerned with the problems of the rural community.

When first he arrived on the diamond fields the river diggings had been largely abandoned; his new constituents were, for the most part, Afrikaans farmers—hardly promising recruits for his Imperial campaign. It may have been an atavistic urge that attracted him to the farming electorate—'My ancestors', he was fond of saying, 'were keepers of cows'—or he might simply have welcomed the chance of an easy, inexpensive election. Whatever his motives, his choice was wise. Although he was not opposed at Barkly West in 1881, he was later to hold the seat against fierce competition. Even when his reputation in South Africa was at its lowest, nothing could dislodge him from his entrenched position. He represented Barkly West until the day he died.

The first candidates to offer themselves for the Kimberley seats were two former members of the old Legislative Council: Dr Josiah Mathews, a popular, cultivated and relatively liberal physician, and that indefatigable champion of the Good Templars, Mr George Bottomley. Several other names were bandied about as possible contestants. One of the most frequently mentioned was, of course, J. B. Robinson. The former mayor's political ambitions were well known; as he had not stood for re-election to the Town Council, it was naturally assumed that he would stand for parliament. To everyone's atonishment, he declined nomination. At the beginning of 1881, he let it be known that he had no intention of standing. He was, he claimed, fully occupied with his newly formed mining companies and it was quite possible he would be away from Kimberley on nomination day.

Robinson's refusal came as a blow to the mining fraternity. Both Dr Mathews and George Bottomley had mining interests but they were by no means as involved, or as exclusively concerned with the diamond trade as J. B. Robinson. In any case, Dr Mathews' liberal views—coupled with his disastrous reputation as a gambler—and George Bottomley's teetotalism did not please all the financiers.

In an attempt to find a more representative candidate, Barney Barnato was asked to take Robinson's place. Barney had also resigned from the Town Council and appears to have been ready to stand, but not everyone regarded him as a suitable choice. He was far too flashy. His theatrical connections smacked of flippancy and both his private, and his business, life left much to be desired. The thought of the flamboyant Barnato representing Kimberley's financial interests in the staid Cape parliament must have horrified many a serious-minded capitalist.

As it happened, Barney's candidature was not put to the test. At the last moment Robinson was prodded into accepting nomination. It was explained that as Robinson had never been requisitioned he had at no time officially withdrawn. 'Yesterday, however,' explained the *Independent* on 18 February 1881, 'several influential gentlemen saw Mr Robinson upon the matter, and, at their urgent entreaty he has consented to reconsider his decision and allow himself to be put in nomination.'

There was no lack of enthusiasm about Robinson's election campaign, however. The requisition asking him to stand was signed by several financiers—including Cecil Rhodes and Barney Barnato—and no expense was spared in ensuring his election. The *Independent*, of course, gave him full backing and Barney Barnato became his campaign manager. Barney, more at home on a public platform than on a parliamentary bench, was in his element. He took the chair at most of Robinson's meetings and spoke at such length that nobody else got a chance.

A week after Robinson's candidature was announced, for instance, Barney addressed three hundred Robinson supporters at one of the largest meetings of the campaign. 'Mr Barnato spoke ably and well,' it was reported, 'expatiating upon the views Mr Robinson had expressed upon all important questions, and entering into detail on the reasons which should influence the digging and diamond buying interests to support him as their representative. Mr Barnato spoke for nearly an hour, and was, throughout, listened to with the most eager attention. The meeting concluded at a late hour, with a hearty vote of thanks to the indefatigable and universally popular chairman and a unanimous vote of confidence in Mr J. B. Robinson.'

Election day was on 15 March. There was little doubt about the outcome. Although a fourth candidate had put in a fleeting appearance, he had soon dropped out of the running. Of the three remaining contenders—Mathews, Bottomley and Robinson—the abstemious Mr Bottomley was more or less doomed from the start. The fight between Dr Mathews and J. B. Robinson, however, was both intense and hectic. It reached a climax the evening before polling day. 'Mr Robinson's supporters', reported the biased *Independent*, 'perambulated the town headed by a capital brass band, and were addressed at several points by Mr B. Barnato and other gentlemen. A motley rabble, carrying flaring rags dipt in paraffin, and yelling like so many demons represented Dr Mathews, and distinguished themselves by pelting the supporters of Mr Robinson with stones and causing several injuries.'

The 'rabble' evidently represented the majority of voters, for Dr Mathews finished top of the poll, with Robinson coming a close second, and Bottomley a poor third. Robinson's triumph at being elected was somewhat marred by his having to take second place. He was now the 'Junior Member for Kimberley'.

[4]

Rhodes, who had been unable to express his views in the election, made up for lost time once he took his seat in the Cape Assembly. He was the first of the new members from Griqualand West to make a speech. The reaction to it was mixed. To some, his awkward gestures, his nervous high-pitched voice, unkempt appearance and complete disregard for parliamentary etiquette (the Speaker had repeatedly to remind him to address members by their constituencies, not their names) were hardly compensated for by his obvious enthusiasm. He was considered too impetuous to make a lasting impression.

'I remember his first appearance in the House,' said a political journalist: 'a fine ruddy Englishman, a jovial-looking young squire. His speech was bluff and untutored in style, with no grace of oratory. A candid friend remarked afterwards that he would be a Parliamentary failure.' Others were not so sure. The more discerning appreciated the force of his arguments and recognised the cool, if unspectacular, reasoning of his badly delivered speech. 'I heard several members as well as strangers in the Gallery', wrote a Kimberley reporter, 'speak highly of his maiden effort, from which they predicted he would make his mark in a quiet unassuming way. In his hands, at least, the dignity of the House will not be lowered.' This was more encouraging, but it was still not saying much. Quite obviously Cecil Rhodes was not expected to set the continent on fire.

More was expected from J. B. Robinson. Unlike Rhodes, Robinson was well-known in the Cape. Not only was he recognised as one of the richest men on the diamond fields, but his periodic outbursts, his continual litigation and his recent term as Mayor of Kimberley, had kept his name well to the fore. When he rose to speak, three days after Rhodes, the benches on both sides of the assembly were packed. He acquitted himself well. He was an experienced speaker and, meeting with no opposition, was in full control of himself. There was no need to hold a post-mortem on his debut. 'Mr Robinson was received with loud cheers, by a crowded house,' was all that needed to be said.

The first parliamentary session attended by the Griqualand West contingent was taken up, largely, with discussions on the conflict between the Cape and the Basutos. The conflict had been greatly aggravated by the decision of the Cape Prime Minister, Gordon Sprigg, to disarm the Basutos. The disarmament question was one of concern to the diamond industry. Africans were attracted to the diggings mainly by the desire to buy firearms. All the candidates in the Kimberley elections had made their stand quite clear on this issue.

Robinson had firmly supported the Cape government. He recognised the need to attract cheap labour but, having fought against the Basutos as a young man, he had no intention of being soft on the 'native question'. 'I am no Negrophilist,' he had stated, 'though I am in favour of the natives being treated with justice and humanity.' He upheld the racial laws passed by the old Legislative Council and emphasised that the 'natives' must be made industrious and compelled to follow peaceful pursuits. 'This desirable object', he said, 'is not incompatible with humane and firm treatment, and one of the first lessons to be instilled into them will be to respect the laws of *meum* and *tuum*. The Diamond Fields have, in my opinion, done much to accomplish this.'

His was the voice of the British racial paternalist; to be heard throughout nineteenth-century Africa. Under the guise of material and spiritual uplift, the African must be made to work and know his place. It was a philosophy which struck deep roots in English-speaking South Africa and has flourished well into the second half of the twentieth century. In 1881, Robinson's racialism could at least be attributed to the age in which he lived.

He was vehemently opposed by Rhodes and Dr Mathews. From the outset, Dr Mathews had let it be known that he considered the treatment of Africans at the diggings to be scandalous. He had been attacked on this issue from all sides, but he had stood firm and held to his view in the Cape parliament. Rhodes's attitude was—as always on such matters—more equivocal. One of his biographers has said: 'Rhodes and the Kimberley people disliked the [Basuto] war, because they had many Basutos working for them in the mines and as one man said, "After all, we sold them the guns; they bought them out of their hard-earned wages, and it *is* hard lines to make them give them up again." ' Whether this was Rhodes's view, or whether he was solely concerned with the labour force, is not certain. In his election manifesto he had favoured the Basutos retaining their arms and had stood out against the confiscation of Basuto territory. On the other hand he had advocated that the power of the Basuto chieftain be broken. How far he was therefore committed to support the Cape Government was arguable. The argument was settled soon after he took his seat in the Assembly.

'It is generally admitted', wrote a Kimberley reporter, 'that the little group of four from Griqualand West will always be a force in the House.' The significance of this remark lay in the fact that the new members more or less held the balance of power between the Government and the Opposition. The Prime Minister's majority was precarious; when Rhodes and Mathews voted against him on the Basuto question they contributed in no small measure to his downfall. They also made a lasting enemy of J. B. Robinson.

At the end of the parliamentary session, the representatives for Griqualand West were welcomed back to Kimberley at a banquet given by the Mayor. Rhodes's colleague, Frank Orpen, could not attend but he was not missed. Throughout the session attention had been focused on Robinson, Rhodes and Mathews; everyone was anxious for them to give an account of themselves. The after-dinner speeches were expected to be lively. They were.

Dr Mathews, as senior member for Kimberley, spoke first. He was listened to attentively and politely applauded. J. B. Robinson, who followed, was greeted with loud cheers. He more than lived up to his peppery reputation. In a passionate attack, he denounced Rhodes and Mathews for contributing to the collapse of the Sprigg administration. He accused Rhodes of cynically breaking his election promises.

So inflammatory was his speech that when Rhodes rose to reply he was 'received with cries of "rat" and mingled cheers and hisses'. This was Rhodes's first experience of facing a hostile audience: he stood up to it well. After a long reasoned explanation of why he had voted against Sprigg, he contemptuously dismissed Robinson's accusation that he had 'ratted' on the electors. He sat down to loud cheers.

Robinson immediately jumped up to deny that he had called Rhodes a rat, but he was cut short by a Mr Neale who brought the proceedings to a close with a rendering of 'The Village Blacksmith'. However, it needed more than a song to disguise the fact that two of the most important men in Kimberley were now declared enemies.

There was no chance of the quarrel blowing over. In fact, within months, things became a great deal worse. In September 1881, J. B. Robinson was involved in yet another libel action against the *Diamond News*. Technically he won the case, but, instead of the large sum he had sued for, he was awarded a mere 50s. damages. His humiliation was all the more galling due to the fact that important evidence had been given against him by Cecil Rhodes. Robinson was a bad loser and a 'good hater'. To the end of his life, it is said, he 'hated Rhodes with deep hatred'.

PART TWO
THE MINERS

CHAPTER TEN

'*A SIGHT WORTH SEEING*'

BARNEY BARNATO's former partner, Lou Cohen, was nothing if not versatile. A dealer in diamonds, journalist, raconteur, and amateur actor, he was, it is said 'everything by turns and nothing long'. The optimistic Lou Cohen was only too ready to offer his services to any promising venture. Thus it was that, sometime in 1880, he accepted the post of manager to a theatrical troupe touring the Transvaal. He was attracted to the job—as he gleefully admits— more by the prospect of seducing one of the chorus girls than by the promise of a new career. When the show closed in Pretoria a few months later, Lou found himself out of work and out of love.

Nothing daunted, he made his way to Cape Town where, in an unexpected burst of patriotism, he volunteered to join a British contingent fighting in Basutoland. His brief military career proved exciting, but not particularly glorious. It ended when he broke his ribs falling from a Cape-cart. As soon as he had recovered, he decided to forgo the delights of love and war and return to his more predictable calling as a diamond dealer. He had been away from Kimberley for over a year. Arriving back in the town, he was startled by the developments that had taken place during his absence.

'I found the place strangely altered,' he says. 'The old time digger, the farmer-digger, the gentleman-digger had almost disappeared, and in their place had sprung up a mushroom breed of financiers, who were destined in the near future to put their hands deep in the pockets of the British public, and form the cradle of a brood of costers and aliens, whose business methods later on made South African company promoting a vehicle for wholesale plunder and chicanery. In 1881, that bubble year, there were in Kimberley more than a dozen companies.'

But, if business was being conducted on a different scale, the town itself was much the same. It was still as hot, still as dusty, still as plagued with flies and still as unlovely as ever. The majority of its buildings were still mere corrugated iron sheds; one could almost count the straggling, dry-leaved trees. The bars were packed with bragging customers, the shops crammed with expensive goods; hopeful traders were still dreaming of the day when the arrival of the railway would solve their transport problems. As yet, all goods had to be transported by ox-wagon from the nearest railhead and all passengers obliged to travel by Cobb's famous, but uncomfortable, coaches. The complaints about the drainage, the water supply, the pot-holed roads and the local politicians were as loud as ever. In the jostling streets, it was still possible to spot familiar faces.

Conspicuous among the older inhabitants was the decidedly more prosperous but just as matey Barney Barnato. Always a dandy, Barney was now in a position to indulge his passion for clothes to the full. He always looked immaculate. The heat might be gruelling but with his pince-nez, his starched collars, his bow-ties, his button-holes, his checked suits and his spats, he cut a jaunty figure. Special occasions would see him more splendidly turned out—in top hat, cravat, watch chain and morning suit.

In common with the majority of Kimberley's rich men, Barney had not bothered to build himself a Kimberley home worthy of his wealth and status. None of them planned to live out their lives in this dust-blown, stifling and graceless town. 'Even people doing well with their diamonds', wrote one vistor, 'live in comfortless houses, always meaning to pack up and run after this year, or next year, or perhaps the year after next.' The idea was to make one's money and get out. Rhodes was one day to live among the Cape-Dutch magnificience of Groote Schuur; Robinson amongst the splendours of Dudley House in London; Barney Barnato was to build himself a lavish, pseudo-Gothic mansion in Park Lane. It was with good reason that Trollope had asked 'Why try to enjoy life here, this wretched life, when so soon there is a life coming which is to be so infinitely better?' That theory had long consoled Christians, said Trollope, so why not the rich men of Kimberley?

Barney would have needed more than a Kimberley mansion, however, to impress the jaundiced Lou Cohen. 'If Kimberley had altered, Barnato had not,' he said, 'except he dressed better, and had pushed himself into circles which a year ago would have none of him, and whose unsuspecting members were ultimately glad enough to bow to his cunning, and scramble for the crumbs that fell from his table. Insistent, persevering, unscrupulous, his success made him dominant, not just or generous. All the same he still promenaded nightly, with nothing in his pocket but an annexed billiard chalk, never failing as cueist or card player to make his weekly expenses. People of his own kidney . . . followed him blindly as a Hero of Enterprise and Tradition, with the consequence that all they possessed—and in many instances they had amassed considerable amounts —helped to swell that baneful treasure which rolling snowball fashion increased in bulk.'

Had Lou Cohen been fortunate enough to be a shareholder in one of the Barnato companies, he would undoubtedly have viewed his old friend's prosperity differently. What he has to say about Barney must be treated with caution. However, prejudiced as it is, his assessment of Barney's new status certainly reflects the rapid strides that had been made by the Barnato brothers.

They could not—or so it appeared—put a foot wrong. Where others struggled with landfalls, faulty machinery and flooded claims, nothing seemed to interfere with the diamond yield from the Barnato properties. Even hardened diggers, used to the strange quirks of fortune in Kimberley, marvelled at the ease with which Barney and Harry Barnato seemed to better their rivals. Shortly after launching their companies, they opened new central offices in Natal Street and there displayed their diamonds. The show was one of the wonders of Kimberley. 'The Barnato Company's output since they commenced operations is a sight worth seeing,' reported the *Independent* in June 1881. 'The stones fill a fair-sized cashbox and among them are some of the most perfect gems we have ever seen.' The following month the brothers published their first quarterly report which showed that they had mined 10,328¼ carats of diamonds in three months and sold them for £17,478 6s. 3d. In addition to this they had realised a further £600 by the sale of 1,274 carats of fine sand stones, bringing the gross yield from their ground to £18,078 6s. 3d.

This result was made all the more remarkable by the fact that it had been achieved with minimal cost to the company. 'The working expenses, thanks to good management,' said an admiring financial columnist, 'were exceptionally low, lower in fact than we remember to have seen them in the case of any other company in Kimberley, amounting to but £5,198 for the quarter. The directors have declared a dividend at the rate of 9 per cent, for the quarter, and shareholders have every reason to expect an even more satisfactory return for the next three months.'

Elsewhere in Kimberley the story was by no means as cheerful. When J. B. Robinson's Standard Company published its second interim report a few days later, for instance, a very different picture emerged. 'The Company has been worked under great, but, it seems, wholly unavoidable disadvantages during the last quarter,' it was said. 'Hauling blue [ground] was only possible during about half the period . . . vexatious delays caused by the falls of reef and heavy rains. . . . The Directors do not feel justified in declaring a larger dividend than 4 per cent for the last quarter.' The extraordinary thing was that Robinson's Standard Company ('well-known to comprise some of the richest claims in Kimberley') should have suffered so much from rain and falling reef, while the adjoining claims of the Barnato Company had seemingly escaped these natural disasters.

Barnato Company shareholders had no doubt that the result had been achieved by the indefinable Barnato genius. At a general meeting, presided over by the ebullient Barney, they applauded loudly when they were assured that the first quarterly report was not a freak return. 'There is every prospect of a much larger dividend next quarter,' Barney told them, 'as the Company would not have so many drawbacks to contend with. They had had no flukes during the past quarter, and the fall in the diamond market at home had seriously affected them. . . . The stock was all in local hands and would not be depreciated by any external influence.' It was all most encouraging. Barney enjoyed himself immensely acting out the role of a successful company chairman. He played to the gallery and was duly encored.

The shareholders, however, were all Kimberley men and were not completely blinded by his performance. Most of them had known the Barnato brothers since the early days and were well aware that, while Barney tended to hog the limelight, he was not solely responsible for the company's success. Their sentiments were summed up by Dr William Murphy (J. B. Robinson's friend) who, at the end of the meeting, proposed a special vote of thanks 'to Mr H. Barnato, whose energy and perseverance had been mainly instrumental in the successful working of the Company'. The motion was passed unanimously.

[2]

Less biased observers would undoubtedly have questioned the nature of Harry Barnato's 'energy and perseverance'. The rumour that the Barnato brothers established their fortune by illicit diamond buying was to dog them throughout their careers. This is not surprising. Few men dealing in diamonds escaped such accusations. Visitors to Kimberley were astonished at the way it was openly admitted that vast fortunes had been accumulated illegally. 'As for the moneyed men on the Fields,' wrote one such visitor, 'is it a libel to say that most of them owe their wealth either to illicit diamond buying or to taking advantage of the necessities or inexperience of unfortunate diggers? . . . Some of the most prominent men of the place were yesterday selling umbrellas in the streets of London, or catching birds on Hampstead Heath. And yet, although everybody knows all this, everybody winks at it.'

Those caught trafficking in stolen diamonds were sentenced to imprisonment or, in more serious cases, to a term of hard labour on the breakwater at Cape Town. The guilty included every strata of Kimberley society—from Louis Hond, one of the men who had confirmed the value of the famous 'Eureka' diamond, to an African known as 'Bloody Fool' whose deadly assault on his wife with a crowbar was considered secondary to his I.D.B. offence. When Lady Florence Dixie, a spirited English journalist, was taken on a tour of the Kimberley gaol in 1881, she noted: 'Every species of malefactor seemed to be gathered together. There were a great many in for diamond stealing and illicit diamond buying, men as well as women, some of whom were respectable people of the upper class in Kimberley.'

Even so, the majority of those caught and convicted were minor offenders; men and women who dealt in two or three diamonds at a time. It was well known that I.D.B. was often organised on a much larger scale. The failure of the police to trap the men behind such large-scale operations had resulted in a general lack of confidence in police methods; ways were sought to tackle the problem unofficially. In March 1880, for instance, the Kimberley, Dutoitspan and De Beers Mining Boards held a joint meeting to tackle the possibility of financing a private detective system. This immediately raised a number of delicate questions. In his opening address, the chairman wanted to know if the meeting 'had for its object the detection of this

[I.D.B.] traffic, whether amongst licensed diamond buyers, brokers or diggers?' Everyone was well aware of what he meant: an efficient detective system was bound to uncover a good deal more than petty pilfering by African labourers. It was quite possible that a number of prominent citizens would be implicated.

J. B. Robinson, who was very active in the fight against I.D.B., underlined what they were all thinking. He told them that the British Government had recently sent out an officer to enquire into the official detective system in Kimberley. 'I had a long discussion with him on the subject,' said Robinson; 'as I understood him his view was this: "Your detective system as worked at present is not efficient, as I find from some enquiries that only the small ones are caught here and the large ones escape," and he said "a system like that isn't in force anywhere in the world".' Robinson proposed that they should act on the officer's advice and employ special detectives from Scotland Yard which, he said, would cost them about £12,000 a year.

One of the important results of this meeting was the establishment, on 15 June 1881, of the Diamond Mining Protection Society in which J. B. Robinson played a prominent part. The object of the society was to agitate for an amendment of the laws governing the diamond trade and to assist the inadequate detective force then operating in Kimberley.

But even the sharpest detectives were unable to make much headway. Unless criminals could be caught red-handed in an illicit diamond transaction, it was extremely difficult to provide the evidence necessary for a conviction. To track down large quantities of stolen stones was virtually impossible. Claimholders did not have to account for the stones they sold: it could be assumed that stones in their possession had been found in their claims. The more claims a digger worked, the more safely could he dispose of large quantities of diamonds. If a claimholder was also a licensed diamond dealer, the disposal of stones—wherever they came from—was a simple matter indeed. 'The licensed diamond buyers,' says a digger, 'although sometimes run in, yet too often bamboozled the detectives on account of their being duly licensed buyers of diamonds, when they shielded behind their licence.'

The Barnato brothers owned several claims and were influential diamond merchants. The exceptional yield from their ground, unimpeded by bad weather and obtained at the lowest working costs in Kimberley was bound to renew suspicions. It was not without significance that many of those who had associated with Harry and Barney in the days of the notorious London Hotel were well known to be involved in the I.D.B. trade.

Suspicion was one thing, proof another. Magnificent stones continued to be found in the Barnato claims, but how they got there remained a matter for speculation. Such speculation was as endless as it was hilarious. Many years later, a London racing scandal-sheet, *The Winning Post*, discussed the I.D.B. problem in a scurrilous article on Harry Barnato. 'The Kaffirs were bribed to swallow the "booty" before leaving the mines,' it said, 'a meeting place was arranged, but in what circumstances they passed on the "precious" stones to the purchaser of stolen property history leaves us to conjecture. It would not be incorrect to assume that many of the sparkling oraments which at this moment adorn the neck of a beauty have been subjected to this procedure. If they could speak! Aye, if they could speak. Or if they could cast one bright gleam of their fire which would revivify those meeting places. They probably once reflected them. This would clear the air as lightning clears the skies of foul atmosphere. Many would welcome this miracle.'

[3]

For the moment, however, none of the stories that were told about the Barnatos bothered them unduly. Most of what was being said had been said for years; it had not interfered with their progress. They were doing very nicely. The tales told behind their backs made no difference to the respect with which they were now treated by the Kimberley elite. No longer was it necessary for them to drop their adopted name; Henry and Barnett Isaacs could be forgotten, it was Harry and Barney Barnato who mattered.

This did not mean that they had turned their backs upon their past, or upon their family. Far from it. They sent money home regularly and Barney did his utmost to entice members of the family to settle in South Africa. Surprisingly, he met with little success until his sister Kate Joel was eventually persuaded to send her second son, Woolf, to join his uncles in Kimberley. Barney was devoted to the young Joel boys. They regarded him, it is said, 'as a second father'. Woolf Joel, a quiet, good-looking lad proved an undoubted asset to the firm of Barnato Brothers. He was only fifteen when he arrived in South Africa but he had a good head for figures and, despite a tendency to dark moods, possessed a confidence far beyond his years. After a brief apprenticeship in the sorting sheds, where he quickly picked up the rudiments of the diamond trade, he graduated to office work. Soon he was a recognised favourite among the diamond brokers. 'I doubt', said Lou Cohen, 'if any young fellow was better liked on the Diamond Fields than equable, amiable Woolfie in his youth . . . it was a pleasure to do business with this bright-hearted boy.'

Before long, Woolf Joel was able to report home that he was earning the princely salary of £50 a week, plus a generous commission on the deals he pulled off. His success, and a brief visit to London—dressed in splendid new clothes and handing out money left, right and centre—made his two brothers understandably envious. So much so, that when Harry and Barney returned from their European trip in 1880 they brought Isaac and Solly Joel with them. Barnato Brothers had become very much a family concern.

The business was not confined to diamonds. Once the mining companies had been successfully launched, the Barnatos extended their activities. In July 1881, the Barnato Stock Exchange was founded. The new status enjoyed by Harry and Barney was reflected in the formalities which accompanied the laying of the foundation stone for this new Exchange. The ceremony was performed by the Mayor with a specially made silver trowel—'a very handsome and tasteful piece of work'—and was watched by a huge crowd which included several legal and political dignitaries. 'After the laying of the stone,' it was reported, 'the Company adjourned to the present offices of Messrs Barnato Brothers and partook of champagne and more solid refreshments.' The building was to cost £10,000.

The following month, the brothers again left for Europe. Previously their comings and goings had passed without comment. Now their every movement was reported with deference. 'Among the passengers by yesterday's coach to Capetown', noted the *Independent*, on 13 August 1881, 'were Messrs B. and H. Barnato, who sail for Europe by the *Pretoria*. "Barney" (as he is familiarly called by his friends) bids fair to develop into a second Baron Albert Grant when he once gets a footing among the City leviathans—that is to say, if he doesn't attempt to wrest the laurels from Henry Irving in another sphere. Shall we ever forget our Mathias, or the noble signor who was wont to show us his proficiency in the art of legerdemain?

Never. Yet we think these brothers have achieved their greatest triumphs in the financial world, and cordially wish them *bon voyage.*'

Even among Kimberley's *nouveaux riches*, the achievements of the Barnato brothers were regarded as truly outstanding. Starting as impoverished hucksters, with little to recommend them other than their entertainment value, they had displayed a financial talent which, on the surface at any rate, seemed to border on genius. They had proved themselves to be the equals, if not the superiors, of the shrewdest minds in Kimberley. The magnitude of their success far outweighed any consideration of how it had been accomplished. Obviously there were lessons to be learned from their example. Three days after Harry and Barney had departed, the *Independent* was still pondering the Barnato phenomenon. 'Few men', it observed, 'have been so successful within a comparatively short period as they. Arriving here seven or eight years ago, without fifty pounds in hard cash to their names, but blessed with great business capacity and foresight, they have succeeded in amassing a handsome fortune by the exercise of their mental ability, and dogged determination to hold on to what, at one time, certainly appeared little better than a sinking ship. That they have thoroughly deserved this success no-one will be found to deny. Men who can hold on as they did, and then, at the right moment, launch out into the formation of legitimate companies successfully must be gifted with an enormous amount of brain and foresight. Mr B. Barnato is Chairman of one company, and Director of four or five others and Mr H. Barnato has done more perhaps to contribute to the success of their undertakings by the painstaking manner in which he has constantly worked to protect the interests of the shareholders. We wish them both a pleasant trip to the old country.'

CHAPTER ELEVEN

A NEW FRIEND AND A NEW LAW

BEFORE taking his seat in the Cape Assembly, Cecil Rhodes had strengthened his position in the De Beers Mine. This was his first important step towards gaining control of the entire mine.

At the beginning of 1881, the chief rival to Rhodes's newly launched De Beers Mining Company was the firm of Stow, English and Compton—a group of diggers who had been working in the mine since 1878. According to the head of this firm, F. S. Philipson-Stow, a well-known Kimberley lawyer, his company was much better placed than the Rhodes combination. By February 1881, he says, they had 'secured the key to the De Beers Mine. We had succeeded in cutting off the firm of Rudd, Rhodes, Graham, Alderson and Dunmore from the East and West. Besides holding this last position strategically our claims were among the richest.'

Faced with this, his first real challenge in the mine, Rhodes brought his powers of persuasion to bear. 'I have never', he once claimed, 'met anyone in my life whom it was not as easy to deal with as to fight.' This is an oversimplification. He was not always so fortunate in his dealings. But he did deal successfully with Philipson-Stow. Exactly how this deal was accomplished is not certain. Unfortunately, the only record of it is that written many years later by Philipson-Stow. And by that time Stow was too disillusioned about Rhodes's methods to elaborate on them. All he says is: 'When Mr Rhodes realised the strength of the position acquired by Stow, English & Compton . . . he approached me and made overtures which eventuated in our accepting a portion of his firm's holdings as of sufficient value to justify an amalgamation of the two interests.' The final negotiations for this merger were completed in March 1881.

With the acquisition of the valuable Stow, English and Compton claims, it became necessary to put the De Beers Mining Company on a more businesslike footing. For the first year of its existence it had been managed somewhat haphazardly. There was no regular company secretary and, more often than not, Rhodes had had to attend to the firm's correspondence himself. One of the first cheques issued by the De Beers Mining Company, was drawn by Rhodes in his own favour for £5 'as an advance against his salary as secretary'. Such a happy-go lucky state of affairs was no longer possible. Not only had the company's interests expanded but Rhodes was now a Member of Parliament and was away from Kimberley for long periods. It became necessary for the company to engage a full-time secretary. The young man chosen for the job was Neville Ernest Pickering, son of a Port Elizabeth parson.

It is not known whether Rhodes had met young Pickering before he was employed as secretary of the De Beers Mining Company. He might have done. Before coming to Kimberley, Pickering had been employed by the Port Elizabeth firm of Dunell Ebden who, at one time, had owned the Vooruitzicht farm upon which the Kimberley and the De Beers mines were discovered. Although the farm had been sold to the Cape Government for £100,000 in the year that Pickering joined Dunell Ebden, the firm had retained some of its interests on the diamond fields and these interests might have brought Rhodes and Pickering together. However, their acquaintance could only have been slight. This makes their later association all the more remarkable. For, not only did Neville Pickering become Rhodes's closest friend but, within a few months of his taking up the De Beers appointment, the two of them were sharing a small corrugated iron cottage facing the Kimberley Cricket Ground. Such an intimate and rapidly developed friendship was—as far as Rhodes was concerned—quite without precedent.

Rhodes did not make friends easily. His early years in Kimberley had, for the most part, been lonely. He had many acquaintances and was known at times to take part in the social life of the town but he was never very close to anyone. He gave the impression of being aloof, of wanting to keep people at a distance. Even his partnership with Charles Rudd lacked the warmth of an intimate relationship; they were business associates first and friends incidentally. He had found that he had more in common with some of the young men he met at Oxford. Friends that he made at University—men like Sir Charles Metcalfe and Rochfort Maguire—were later to join his African enterprises and to remain loyal to him throughout his life. But these friends were members of a group of undergraduates with whom Rhodes mixed; he never shared rooms with them, nor did he appear to prefer one to the others.

And if he had no close male friends, his name was certainly never linked with that of a woman. The very idea of an intimate relationship with a female seemed to frighten him. He could not understand why any man should want to get married. His objections to marriage were to become notorious in later years, but they were apparent even when he was a young man. Writing to an acquaintance in Kimberley in 1876 (when he was twenty-three) he says: 'I hope you won't get married. I hate people getting married. They simply become machines and have no ideas beyond their respective spouses and offspring. I suppose ———— is married. He is, I think, the most unsatisfactory fellow I ever came across.' Sex played no part in the stories told about Rhodes's early life in Kimberley. If he went to a camp dance, he always chose the plainest girl in the room as his partner and would tell her bluntly that he danced only for exercise. Even Lou Cohen, who delighted in scandalous innuendo, was hard put to question Rhodes's chastity. 'The silent, self-contained Cecil John Rhodes,' he says, 'I have many times seen him in the Main Street, dressed in white flannels, leaning moodily with his hands in his pockets against a street wall. He hardly ever had a companion, seemingly took no interest in anything but his thoughts, and I do not believe if a flock of the most adorable women passed through the street he would go across the road to see them.'

The only people he was known to pursue, were those who could help further his schemes. Neville Pickering hardly fell into this category. He was a bright, efficient young man and he became an excellent secretary for the De Beers Mining Company, but his services were not indispensable. His attraction for Rhodes seems to have been entirely personal. He was all that Rhodes was not. Fresh-faced and sunny natured, he was gregarious, popular, and a great favourite with the girls. Unlikely as the friendship at first appeared, it was to prove enduring. There was obviously something about Neville Pickering that satisfied an unconscious need in Rhodes's complex nature. 'They shared the same office and the same dwelling house,' says Ian Colvin, 'worked together, played together, rode together, shot together.' The more romantic have likened them to David and Jonathan.

Their cottage was very much a bachelor's establishment. It was a place of wooden chairs, bare tables, iron bedsteads and horsehair mattresses; it contained the necessities but made no concessions to comfort. Rhodes, it is said, often slept with a leather Gladstone-bag for a pillow. They were looked after by a coloured man-servant who acted as their housekeeper, cook, valet and even, on occasion, barber. They did little, if any, entertaining at home. The growing snobberies of Kimberley bothered them not at all.

There was no need for Rhodes to live such a Spartan life. He was a very rich man. His income at this time is estimated to have been in the region of £50,000 a year. Whatever conveniences Kimberley could offer were available to him. It was not meanness but complete indifference to his surroundings that made Rhodes live as he did. He would no more have thought of embellishing his house than he would of dandifying himself. The diamond rings, pins and cuff-links sported by other Kimberley magnates—even the austere J. B. Robinson wore a diamond ring—he regarded with distaste. Jewellery on men was, for him, a sign of effeminacy. He dressed as plainly as he lived. Happiest in a shapeless sports coat and baggy white flannels, he looked more like a down-at-heel digger than a prosperous mine owner. Neville Pickering undoubtedly brought warmth to Rhodes's life but he could not change the habits of a lifetime.

Rhodes had greater faith in Pickering than in any other man. Shortly after they moved into their cottage, Rhodes made his third will. He left his entire fortune to his young friend. He handed the will to Pickering in an evelope with a covering note. 'Open the enclosed after my death,' the note read. 'There is an old will of mine with Graham, whose conditions are very curious, and can only be carried out by a trustworthy person, and I consider you one.'

The 'curious conditions' of that old will were, of course, those which provided for the extension of the British Empire by the recovery of the United States for Britain and the colonising of the world by British settlers. What the amiable Pickering would have made of such a legacy, one does not know. He was never put to the test. And Rhodes never placed such a trust in the sole keeping of any other person.

[2]

It must have been shortly after Pickering was engaged as secretary to the De Beers Mining Company that Rhodes left Kimberley for his last term at Oxford. He arrived in England in October 1881 and took his degree in December. It had taken him eight years to graduate. Much of that time had been taken up by his mining concerns in Kimberley; by the time his degree was conferred the reasons for his taking it had largely fallen away.

He had gone to Oxford because he thought a degree would help in whatever profession he might follow. Now there was no longer any question of his having to follow a profession. His career was decided. He was a mining magnate and a politician; his academic qualifications were superflous. But he never regretted his time at Oxford. It had, in his opinion, put him among the elite. 'The Oxford system,' he once said, 'in its most finished form *looks* very unpractical, yet, wherever you turn your eye—except in science—an Oxford man is at the top of the tree.'

The taking of his degree was a mere formality. During his last term at Oxford his mind was on other things. When Philipson-Stow met him in England at the end of the year, he was bubbling over with a scheme he had worked out for amalgamating the four mines of Griqualand West. 'I can see it now,' wrote Stow. 'Folio after folio of intricate figures. . . . The labour bestowed on this project must have been prodigious!' However, Stow was not impressed. He considered that they would be wiser to concentrate on consolidating the various holdings in the De Beers Mine before launching such an ambitious scheme. Rhodes appears to have taken this advice to heart. By the time he arrived back in Kimberley his grand amalgamation plan had been abandoned for a more modest project.

With the help of another of his partners, William Alderson, he had approached Baron Erlanger, the financier, with a proposition for unifying the De Beers Mine. When news of this scheme leaked out, it created great excitement in Kimberley. 'De Beers Mine is the smallest on the Fields and can probably be had for a reasonable figure', said the *Independent*. 'Many a good thing has fallen through before now, simply because the promoters were too precipitate in forwarding it into the market. We hope that mistakes will not be made with Old De Beers.'

Speculation as to the progress being made by the Rhodes group filled the newspapers for weeks. It was rumoured that the amalgamation of De Beers was merely a beginning and that plans were underway for the Kimberley Mine to follow suit. Enthusiasts saw it as the dawn of a new era for the ruinously competitive diamond trade. Rhodes's efforts were supported by encouraging leaders in the capitalist press.

But it came to nothing. Conflicting local interests and suspicion of the foreign promoters made it impossible for the various companies in the De Beers Mine to reach agreement. Just when it looked as if a majority decision was in the offing, some of the companies backed out, demanding a higher valuation of their ground. This infuriated the editor of the *Independent*. 'Owing to the silly selfishness of one or two outside and unimportant companies the scheme for the amalgamation of De Beers Mine seems to hang fire,' he complained. 'It is absurd, however, for these companies to stand out for long prices, since their ground is by no means essential to the carrying out of the scheme. If the promoters will accept our advice they will leave the contumacious outsiders where they are at present—in the cold—or, the reef.'

But Baron Erlanger was not looking for advice. What he wanted was an amalgamation of the entire mine and so long as one or two companies—no matter how small or unimportant—prevented this he was not prepared to back the scheme. Amalgamation was not to be as simple as some seemed to think.

[3]

Rhodes must have foreseen failure. When negotiations finally broke down he was no longer in Kimberley. He had, in fact, already changed his tactics. It had quickly become apparent that to carry through an amalgamation scheme it was first necessary to win the confidence of the mining community. This he did not have. To gain it he would have to establish himself as an undoubted champion of the diamond industry as a whole. An excellent opportunity to do just that was at hand. The Cape Assembly met in March 1882. It was due to debate important legislation affecting the diamond trade. Rhodes was fully prepared to play his part.

In September of the previous year, he and J. B. Robinson had been appointed members of a Mining Commission. The purpose of the Commission was to investigate the state of the diamond industry and to make recommendations for improvements. In effect this meant formulating legislation to control the traffic in illicit diamonds. Not everyone had welcomed Rhodes as a member of the Commission. The chairman of the Commission was J. X. Merriman, a Cape politician who had once tried his luck at the diamond diggings and whose liberal views were regarded with suspicion in Kimberley.

Rhodes's parliamentary alliance with the lanky, intellectual Merriman had made him suspect. 'Mr Rhodes' appointment [to the Mining Commission]', it was said, 'appears to be the one most open to objection . . . everybody knows that so long as Mr Merriman is Chairman of the Commission, Mr Rhodes will follow him as a blind man follows a dog.'

These misgivings were to be proved unnecessary. Rhodes played little part in the deliberations of the Commission. Shortly after his appointment he had departed for his last term at Oxford. It had been left to J. B. Robinson and the Diamond Mining Protection Society to draft the bill that was to be presented in the Cape Assembly. Not only were the provisions of this bill most stringent, but Robinson had succeeded in winning Merriman's full support for the proposed legislation. Merriman had seen it simply as a matter of ensuring honesty and fair dealing in an industry vital to the Cape.

Having lost ground to Robinson during the drafting stages of the bill, Rhodes was determined to make himself heard in the parliamentary debate. He was given plenty of opportunity. The Diamond Trade Act of 1882 had a stormy passage through the Cape Assembly.

The main opposition speaker was James Leonard, the Member for Oudtshoorn and a former Attorney General. He commenced his attack by denouncing the Special Diamond Court; the court that had been set up to deal with I.D.B. offences. Cases coming before this court were heard by three officials—not necessarily magistrates —and the sentences passed were often reduced or quashed on appeal to the Supreme Court. 'The reckless way in which the Court convicted on the most meagre evidence', said Leonard, 'were such as to call for censure from every man who had any regard for right and justice. . . . I have seen cases appealed before the Supreme Court in which convictions have been obtained upon evidence which I am certain no jury in the country would have convicted the commonest black man of the most ordinary crime. . . . The Illicit Diamond Court of Griqualand West is a blot on the judicial system of the country which should not be allowed to remain.'

He went on to condemn the notorious 'trapping system' which was frequently used to obtain convictions. This was probably the most sensitive spot in which to hit those who professed to uphold law and order in Kimberley. The 'trapping system' reflected little credit on anyone. It operated in the most simple and obvious way. An African would be thoroughly searched and then given diamonds to take to a suspect buyer. He would be followed by detectives who, at a given signal, would close in and arrest the buyer in the act of taking the diamonds. Such a system was, of course, wide open to abuse. The African runner was often used purely as an *agent provocateur* and many an otherwise innocent dealer was lured into a police trap before he had time to test the runner's credentials. Stories abounded of men who had been forced into seemingly illicit transactions. It had earned the Kimberley police a bad reputation. Harry Graumann, who arrived at the Cape in the early 1880s, for instance, says that he was afraid to visit the diamond diggings. 'I had been intimidated into believing', he says, 'that anyone was liable to get into trouble at Kimberley in connection with the illicit diamond trade. . . . On all hands one heard stories of detectives putting diamonds into the pockets of innocent people, and then charging them with an offence which has always been punished very severely.' Leonard's attack on the system met with loud approval.

Following him, Rhodes put up a spirited defence of the Special Courts but was somewhat cornered on the trapping system. 'As regard the system of trapping,' reads a report of his speech, 'he admitted that if one was to argue generally that was a system which no-one could approve of. But the circumstances were very exceptional and it was only the great necessities of the community that impelled the sanction of the system. He would like anyone to show him any method or means whereby this crime could be checked except by this exceptional and obnoxious system.' He ended his speech by moving for a Select Committee to enquire into the whole subject of illicit diamond buying. The motion was approved and he was appointed chairman of the Select Committee.

His performance was well received in Kimberley. It was regretted, however, that he had had to launch into what was obviously going to be a tough battle practically unaided. The man to whom everybody looked to defend the diamond industry was J. B. Robinson; but Robinson, at the outset of this crucial debate, was not in Cape Town. He was tied to Kimberley by one of his never-ending libel suits. All he could do, for the time being, was to make the right noises from Griqualand West. And it had become very noisy in Griqualand West.

[4]

J. B. Robinson had sprung into action as soon as reports of Mr Leonard's clash with Cecil Rhodes appeared in the Kimberley papers. That same afternoon he called a public meeting at the office of the Kimberley Mining Board. George Bottomley was in the chair, but Robinson did most of the talking. A huge crowd turned up to hear him.

He quickly let it be known that he considered Rhodes's defence of the trapping system to be feeble. 'Objections have been urged against the trapping system,' he said, 'but it is adopted in England in cases where suspicion of dishonesty exist. . . . Unless you employ this system you might as well hand over your claims to the illicits themselves.' He went on to read out the proposals which he and the Diamond Mining Protection Society had made. These proposals advocated, among other things, that the police should be given power to search suspects without a warrant and that anyone found in possession of a rough diamond, without a permit, should be liable to a minimum of five years' imprisonment. There was also a strong recommendation that the Cape liquor laws should be amended to prohibit the sale of liquor to Africans in Kimberley.

The meeting was not an unqualified success. Robinson was warmly applauded, but it was not long before dissenting voices were heard. There were strong objections to the proposed liquor ban. Canteen owners saw the ban as a threat to their trade and lost no time in letting their views be known. They received considerable support from the commercial sector of Kimberley.

Immediately after the meeting, a whispering campaign began. It was rumoured that plans were underway to herd the Africans into compounds where they would be served by mine-owned shops. There was also talk of an extension of the searching system. Once the Africans were shut up, it was said, the mining companies would tighten their grip. Everyone employed on the mines would be searched before they left work: white overseers, as well as black diggers, would be forced to strip and be subjected to an undignified examination.

A few days later placards attacking Robinson and George Bottomley were posted throughout Kimberley. They called upon mine overseers to attend a mass protest meeting. There was no indication of who was organising this meeting, but it was obviously inspired by the canteen owners. Robinson proved more than a match for them.

On the evening of the meeting several thousand people packed the Market-square. Robinson arrived early. Determined to have it out with his anonymous opponents, he stamped about defying them to appear. For once the crowd—including the mine overseers—was on his side. 'Mr Robinson', it was reported afterwards, 'is a man of far too much pluck to allow professional libellers to trade on their ability to defame him . . . [he] attended the meeting on Wednesday night to meet his accusers face to face, and he never more thoroughly enjoyed the confidence and respect of his constituents.' Unfortunately, having braced himself for a fight, he was rather put out of his stride when his opponents did not show up. There was no sign of the organisers of the meeting. When it looked as if the crowd would disperse for want of a speaker, Robinson and a few others mounted a platform and made rallying speeches. There was no need for them to say much. A petition to the Cape Assembly —opposing trial by jury and the sale of liquor to Africans—was presented and approved and the meeting fizzled out.

'The 19th of April, 1882,' claimed a newspaper, 'will remain a red-letter day in the history of the Diamond Fields, for the reason that on the day named certain infamous combinations against the mining interests were defeated on their own ground, and taught a lesson for all time to come.' This, considering the poor showing of the opposition, was somewhat overstating the case.

The anonymous attacks on J. B. Robinson were to continue for some months. It was said that he had taken over the meeting because he was afraid that his own dealings with illicit diamond buyers would be revealed. Nobody, except Robinson himself, took this accusation seriously. Robinson, of course, threatened to sue for libel but was prevented from doing so because his detractors refused to come out in the open. For once he was obliged to forgo the ultimate legal satisfaction.

In the Cape Assembly the fight went on. Robinson hurried to Cape Town to play his part. He left Kimberley the day after the Market-square fiasco and was in the House of Assembly when the Select Committee, appointed to investigate illicit diamond buying, reported back. Rhodes had been chairman of the committee and its report came as no surprise. The crime of I.D.B. was shown as a formidable threat to the diamond industry and stringent legislation (including 'searching overseers and other employees') was forcefully recommended. Rhodes and Robinson teamed up to push the draconian laws through parliament. Neither of them was particularly happy about the alliance. In politics they were—and continued to be —natural enemies. But the financial threat proved stronger than political considerations and this, together with their determination to outshine each other, ensured their success.

They had to fight for practically every clause of the bill. Most times they won. Only Rhodes's attempt to retain flogging as a punishment for buyers, as well as stealers, of illicit diamonds was rejected outright. Dr Mathews, true to his liberal principles, was largely responsible for this defeat. 'I decidedly objected . . .', he says, 'to flogging being inflicted for what was not a crime against the person but against property. I was so far successful that such brutal ideas were expunged from the act.' Rhodes, however, was to advo- cate these brutal ideas throughout his life.

For his part, Robinson fought the suggestion that three qualified judges should be appointed to hear I.D.B. cases. He had much experience, but little faith, in the courts of Griqualand West. In his opinion, knowledge of the diamond industry counted for more, when dealing with cases involving stolen diamonds, than did knowledge of law. He was only partly successful in his objections. A compromise was reached whereby one judge and 'two others' were appointed to the Special Court.

There were very few other compromises. Most of the recommen- dations of the Diamond Mining Protection Society were adopted without modification. Detectives were authorised to search for rough diamonds without a warrant; all diamonds passing through the hands of dealers had to be registered and a monthly return made; any diamond-buyer suspected of an illicit transaction could be stopped, searched and have his books impounded; the onus of proof of legal possession of diamonds was thrown on the individual in whose custody they were found. The penalty for nearly all the offences listed was 15 years' imprisonment or a fine of £1,000.

The Diamond Trade Act 48 of 1882, as it became, was a remark- able piece of legislation. Visitors to South Africa were constantly shocked by its provisions and by the sentiment that inspired it. 'A law of exceptional rigour punishes illicit diamond buying, known in the slang of South Africa as I.D.B.ism,' wrote Lord Randolph Churchill. 'Under this statute, the ordinary presumption of law in favour of the accused disappears, and an accused person has to prove his innocence in the clearest manner, instead of the accuser having to prove his guilt . . . this tremendous law is in thorough conformity with South African sentiment, which elevates I.D.B.ism almost to the level, if not above the level, of actual homicide.' Others saw it purely as a capitalist charter.

'We had great difficulty in passing the Diamond Trade Act,' Robinson was to say. 'This Bill took weeks and weeks—nearly the whole session of 1882.' But he considered it time well spent. As he grew older, Robinson became more and more proud of his fight on behalf of the diamond industry. He delighted in telling of how he had personally canvassed every member of the Cape Assembly and talked a majority into voting for the act. The laws governing the diamond trade he claimed as his own. Without his efforts, says his biographer, the diamond industry would have collapsed.

Rhodes is rarely, if ever, associated with the Diamond Trade Act. Unlike Robinson, he preferred to forget that he had sponsored this controversial legislation. He had more to boast of in later years and the diamond laws were never universally popular. What perhaps is more important they failed in their purpose. There was a short period—immediately after the passing of the act—when the ships to Europe were said to be crowded with frustrated I.D.B. agents. But new ways of evading the laws were discovered; I.D.B. continued to be a menace. It still is.

However, at the time the legislation was passed, Rhodes posed proudly as its champion. He firmly believed, he said, 'in his baby the Diamond Trade Act, however unpleasant it might be to a portion of the community'. Both he and Robinson were given a hearty welcome when they returned to Kimberley in July 1882. Only Dr Mathews was in disgrace.

Of Rhodes, the *Independent* said: 'On the Diamond Ordinance he spoke with most certain sound, and it is to be regretted that his efforts to perfect the Bill were to a great extent counteracted by the sensational and somewhat imaginative stories of Dr Mathews, notably on the flogging question, which had an undue effect on the House, and compelled Mr Rhodes to withdraw his motion in favour of that useful deterrent.' Dr Mathews could not withstand such criticism. By the time the Cape Assembly met again he had resigned his seat.

There was still one matter outstanding. The new laws did not cover the sale of liquor to Africans. Robinson, cheered on by George

Bottomley and seconded by Rhodes, tackled this issue during the next parliamentary session. He proposed that the Cape liquor laws be amended to prohibit the sale of alcohol to Africans within a five-mile radius of any mine. The amendment was favourably received, but at the third reading of the Liquor Licensing Bill it was unexpectedly defeated by a majority of nine votes. Robinson fumed; the canteen keepers rejoiced.

Kimberley voters rallied round their defeated representatives. Failure proved more rousing than success. At an impressive ceremony in the Kimberley Town Hall, on 25 September 1883, Robinson and Rhodes were presented with an 'influentially signed' address which congratulated them on their efforts on the Liquor Bill and regretted that their amendment had not been accepted. They both made long and pompous acceptance speeches. It was a memorable occasion. Never again were these two rich and powerful magnates to share such popular acclaim. 'By their efforts in the cause of morality and the general well-being of the community,' it was said, 'Messrs Robinson and Rhodes have well earned the mark of regard which is conferred upon them.'

That was one way of putting it.

CHAPTER TWELVE

THE GREAT BARNATO SCANDAL

THE Barnato brothers had kept clear of the illicit diamond buying controversy. Not only was it a delicate matter for them but, for the most part, they were far too busy minding their own more legitimate business. Often they were away from Kimberley for long stretches at a time.

Their frequent absences had given rise to a number of problems. For one thing, Barney had been obliged to resign his position as chairman of their parent company. There was nothing unusual about this. Many of the mining companies in Kimberley had acting chairmen. Sometimes the major shareholders in a company were only indirectly represented on the board of directors. Rhodes directed the affairs of the De Beers Mining Company long before he became chairman. There were periods when Harry Barnato was not even a company director. It caused little surprise therefore when Barney resigned the chair of the Barnato Mining Company. The only strange thing about his resignation was that he allowed a most unsuitable candidate to be elected in his place. The new chairman was J. B. Robinson's friend, Dr William Murphy.

Dr Murphy had been one of the original shareholders of the Barnato Company. He it was who had proposed the gushing vote of thanks to Harry Barnato at the company's first quarterly meeting. However, his admiration for the Barnato brothers had quickly begun to wane. Like his friend, J. B. Robinson, Dr Murphy was a suspicious, touchy, meddlesome man. It was not enough for him to attend the weekly meetings of directors; he had to poke his nose into every corner of the business. Soon he was complaining about the way in which the Barnato Company was run. Certain procedures, particularly the morning 'wash-ups', struck him as highly irregular. He felt duty bound to present his findings to his fellow shareholders. His report did not please the brothers Barnato.

'Wash-ups' were the accepted practice of all mining companies. Although diamond mining had become more mechanical, the basic procedure for recovering diamonds remained simple. Loads of blue ground (kimberlite) were hauled from the mines and transported to the sorting sheds of the various mining companies. Here the ground was spread out, wetted from time to time, and left to disintegrate. A gang of African labourers, watched by a white overseer, helped the disintegrating process by breaking up the larger lumps of earth with picks. Every day a certain amount of the crumbled earth was put through washing machinery. The diamonds recovered from these daily 'wash-ups' were placed in an envelope and sent to the company's main office. The mine manager was responsible for sealing the envelope and, by rights, they should have remained sealed until they were opened at the weekly meetings of the directors.

According to Dr Murphy, this important procedure had been shockingly abused at the Barnato Mining Company. The wash-ups had become very much a Barnato family affair. When he and another director had investigated the way in which the morning wash-ups were conducted, the mine manager had made it plain that their interference was resented. He had therefore found it necessary to draw attention to 'the gross irregularity of the Manager, Mr Pippin, allowing Messrs Woolf and Isaac Joel, the nephews of Messrs Barnato Bros, and simply shareholders in the company, to receive the diamonds at the wash-ups and bring them to the offices of the company at whatsoever time they thought proper'.

In order to remedy, what he called 'this glaring evil', Murphy had ordered the manager to wash-up in the afternoon so that one or more of the directors might attend. The manager had strongly objected. An offer was made to put a cart at Murphy's disposal so that he could, if he wished, be present at the early morning wash-up. When he refused this offer, the matter was allowed to drop. Mr Pippin, the manager, was highly respected. The shareholders preferred to take his word against that of Dr Murphy's.

It is surprising that, after this, the Barnatos allowed Murphy to become chairman of the company. If they hoped to placate him, they failed. Following his election as chairman, in March 1882, the doctor busied himself with the company's affairs to an even greater extent. He was full of complaints. He objected to the fact that there were no proper scales and weights in the office. He resented the manager's habit of exhibiting diamonds to the Barnatos' friends and relations before they had been seen by the directors. His repeated requests that the manager furnish a proper weekly report were ignored. He was suspicious of the fact that the manager received his salary from a number of different sources. 'A mode of payment', he said, 'which I believe does not obtain with any other company or claimholder in the Kimberley Mine.' In fact, he found the entire relationship between the Barnato brothers and their manager highly questionable.

Things came to a head in October 1882. In the middle of the month, Murphy called a special shareholders' meeting. He explained that the friction between himself and Mr Pippin had reached such a pitch that he had no alternative but to ask for the mine manager's resignation. This caused an uproar. Pippin, who was also a shareholder, heatedly defended himself. Half-way through the meeting, the exchanges between Murphy and the mine manager became so abusive that the reporters in the room were requested not to take notes. The proceedings ended with Murphy thumping the table and calling Pippin a liar; the mine manager responded by handing in his resignation.

There was, of course, no question of the Barnato brothers allowing Pippin to resign. If anyone had to go it was the meddlesome doctor. This was made quite clear when, three days later, an advertisement appeared in the Kimberley newspapers calling another meeting of the Barnato Company 'to consider removing the Chairman, Dr Murphy, from the Board of Directors'.

The second meeting, held on 4 November, was conducted with more dignity, if no less animosity. The Barnatos had ensured the outcome. The room was packed with their supporters and they had arranged for the most respected member of their family—their cousin David Harris—to take the chair. Harris seems to have been somewhat bewildered by his sudden elevation to Chairman of the Barnato Company. He had to ask for the meeting's indulgence as 'he was almost ignorant of the terms of the trust deed' and might be mistaken in his rulings.

Dr Murphy, on the other hand, was quite sure of his ground. After he had agreed to accept David Harris as chairman, he was allowed to make a long statement. He read his numerous complaints from a prepared list: starting with the suspicious behaviour of the young Joels and ending with a strange discovery he had made the previous evening. On inspecting the day's findings, he said, he had been amazed to uncover an unrecorded 26 carat diamond. The chairman called upon Mr Pippin to explain this extraordinary occurrence.

Pippin was only too ready to do so. He said that late the previous evening 'Mr H. Barnato came up and asked him if he had had any luck, when he told him that he had found a twenty-seven, which was the weight he judged the stone to be. Barnato asked to let him see it, and he handed it to him. He then said he would play a "lark" on Dr Murphy, and put the diamond in the fine sand.' This explanation went down well. It was typical of Murphy's groundless suspicions and the merry way in which the Barnatos conducted their business. The fact that Harry Barnato had failed to inform the doctor of his deception so that his 'lark' had no immediate point, did not disturb anyone. Murphy had been exposed as a tiresome busybody who was best out of the way. When the resolution to remove the doctor from the Board of Directors was put, it was passed without opposition.

The overwhelming vote against Dr Murphy requires some explanation. What few replies were made to his accusations were flippant and, on the whole, unconvincing. The reason why he was not taken more seriously was due largely to his reputation as a trouble-maker. Few men in Kimberley were as heartily disliked as William Murphy. He had, at one time or another, quarrelled with nearly everybody. Even J. B. Robinson's churlishness was a poor match for the perversity of his friend. Dr Murphy had been known on occasions to beat up Africans, to attack a local chemist, to pull the beard of an elderly club-member and, on one memorable night, to send his own wife screaming into the street. Such behaviour, by Kimberley standards, was not particularly exceptional. Fights could, and did, break out at the drop of a hat. Wife beaters were not unknown. Most men prided themselves on their pugnacity. What made Murphy so objectionable was that his aggressiveness resulted from a very ugly disposition. Prominent among his many unlikeable traits was a virulent anti-semitism.

Murphy was noted for his attacks on Kimberley's Jewish community. He was an unbridled racialist. So Teutonic were his views that he was reluctant to admit that his own father was an Irishman. He tended to blame all his misfortunes on racial 'conspiracies' rather than acknowledge his inadequacies. At the Barnato Company meeting, for instance, he had started his defence by referring to the shareholders as 'the tribe and supporters' of the Barnato brothers. When asked what he meant by 'the tribe' he replied: 'Are you not the tribe, the chosen tribe?' David Harris had done his best to calm the meeting. 'The remark', he said mildly, 'was very irritating. We are not here to discuss religious matters.' But the harm had been done. Few would have minded a reasoned attack on the Barnatos; no-one was prepared to side with a blatant racialist. By his bigoted opinions, Murphy had destroyed his own case.

This was by no means the last confrontation between William Murphy and the Barnato family. The next clash came a year or so later.

[2]

At the beginning of 1884, new elections were held for the Cape Assembly. New seats had been created in Kimberley and Dr Murphy stood as a candidate for one of them. When his name was announced on nomination day, it was greeted with loud howls of disapproval. 'Dr Murphy', it was said, 'will not be the most popular candidate.'

Neither this, nor subsequent demonstrations, discouraged the opinionated doctor. But it was difficult for him to get a hearing. At his first public meeting, the hecklers became so rowdy that Murphy was forced to vacate the platform and leave the hall by a back door. However, Kimberley was nothing if not democratic; the local press came to his rescue. Believing that he had a right to state his views, no matter how unpopular they might be, all three Kimberley papers offered to publish his election address. They soon regretted their generosity.

The editor of the *Independent* did not read Murphy's manifesto until it had been set up and 2,000 copies of the paper printed. When he came to check the lengthy article, however, he ordered all the printed copies to be scrapped and refused to receive any further communication from the doctor. Once again Murphy had given vent to his anti-semitism. His statement, said the *Independent*, was 'a vile and slanderous attack upon a section of the community by which he endeavoured to arouse religious animosity'. Unfortunately, the other two papers were not so scrupulous. They published Murphy's diatribe. By doing so they ensured the doctor's defeat. The Jewish community had done much for Kimberley and was widely respected. It is an indication of that respect that, when the results of the election were announced, Murphy was seen to be bottom of the poll with a paltry 20 votes.

For the Barnatos, the election had an unpleasant sequel. It resulted in the arrest of Isaac Joel. Isaac, the eldest of the Barnato nephews, was, in many ways, the most reckless of the Joel boys. He was twenty-two years old at the time and more prone to trouble than his younger brothers. He and Dr Murphy were old enemies. Isaac had not forgotten the accusations which the doctor had made against him and Woolf Joel. When Murphy published his scurrilous manifesto, Isaac decided to pay off a few old scores.

On the night before the election, he and a few friends had tackled Murphy outside the offices of the *Independent*. A scuffle had taken place: the doctor's hat had been knocked off and Isaac Joel had punched him in the eye. After being rescued by some bystanders, Murphy had rushed to the police and laid charges against three of his assailants—one of those he named being Isaac Joel. But, surprisingly, when the case against Joel came up for hearing the doctor seemed reluctant to press the charge. First, he asked for the action to be postponed and then, at the second hearing, he failed to show up and the case was dismissed.

What happened to make Murphy change his mind is not recorded. He may have been bought off. Neither Barney nor Harry Barnato was in Kimberley at the time but it is possible that their subordinates arranged matters. Certainly Murphy was not the loser. He was taken back onto the Board of the Barnato Company and he and Barney were soon working in close cooperation. This alliance can hardly be regarded as edifying. It is not surprising that the Barnatos were not altogether popular with Kimberley's Jewish community. Lionel Phillips and his wife always referred to Barney as 'that brute Barnato'.

Isaac Joel had been fortunate in his first brush with the law. His next appearance in court was not to end so happily. The following month he was again arrested. This time the charge was more serious. It had to do with the newly passed Diamond Trade Laws.

On Thursday, 20 March 1884, two detectives called at Isaac Joel's diamond buying office in the Barnato Building. Empowered by the new legislation, they demanded to examine Joel's diamond register. Their inspection showed that, of the $1,003\frac{1}{4}$ carats of rough and uncut diamonds on hand, no entry had been made for the receipt of three 10 carat stones. They questioned Isaac about this discrepancy and then asked him to accompany them to the office of the Chief of the Detective Force, John Larkin Fry. Interviewed by the tough, experienced Fry, Isaac became evasive. He explained that the stones had been included in a parcel of diamonds he had once bought, but was unable to specify when, or from whom, the purchase had been made. He was also unable to account for a $16\frac{1}{2}$ carat diamond which he had sold to another diamond buyer ten days earlier. (It had been an investigation into the ownership of this larger diamond that had led the detectives to Joel's office.) After going through the entries in the diamond register and finding no trace of the four suspect stones, Fry put the young man under arrest.

The following day Isaac Joel, licensed diamond buyer, was brought before the Resident Magistrate and charged with a contravention of the Diamond Trade Act. An affidavit put in by the Detective Department detailed two offences: one of selling the $16\frac{1}{2}$ carat diamond and the other of possessing the three 10 carat stones. Bail was allowed and the case was remanded to the Special Court for trial.

As often happened in I.D.B. cases, the trial was subject to many delays. Important witnesses fell conveniently ill and the lawyers found various grounds for postponements. When, on 8 May, judgement was finally given on the first of the two charges—that of selling the $16\frac{1}{2}$ carat stone—Isaac was found not guilty. No evidence had been led; his acquittal was simply the result of a legal technicality. The indictment against him had omitted to state clearly that the stone he was alleged to have sold was not his property at the time. Consequently, the judge had no option but to order his discharge.

The discharge lasted only a matter of minutes. As he stepped down from the dock he was again arrested. The charge against him of illegally possessing three 10 carat diamonds was still outstanding. A fresh bail bond was entered into. This time bail was fixed at £4,000—£2,000 being in his own recognizances and a further £2,000 being put up by his brother Woolf.

Four days later Harry and Barney Barnato arrived back in Kimberley. They had been in England when Isaac was first arrested. When Barney heard what had happened, it was said, 'he packed up his portmanteau and started off for the Cape at once'. He was in a terrible state. As soon as he arrived he rushed to see J. B. Robinson. Robinson had recently been elected chairman of the Diamond Mining Protection Society and Barney was sure that he was in a position to get the case stopped. In the past he had supported Robinson and he evidently saw no reason why he should not ask for a return of favours.

Robinson—no stranger to emotional outbursts—was taken back by Barney's hysterical rantings: as always, when under strain, his persecution complex manifested itself. 'He then explained the whole matter to me, as Chairman of the Diamond Mining Protection Association,' said Robinson, 'and told me, or led me to believe, that he looked upon this action against his nephew as a kind of prosecution against himself, or his firm, and that there was a conspiracy against him. He told me this would have a most serious effect on Joel, and asked me to see Fry about the matter . . . the interview between Barnato and myself took nearly two hours; Barnato was crying most bitterly most of the time. . . . He told me that he was suffering a great deal, tore open his shirt to show me eruptions on his body, and said he could not sleep. It was a most painful scene. He showed me most affectionate letters from his mother, and said he would give £5,000–£10,000, if he could get his nephew clear. . . . I said to him "Do you mean to offer money to me?" and he said "I would give £10,000—I would give it today—to get my nephew free. My nephew is worth £30,000, and can well afford to pay £10,000 to get free." The man was so upset that it was impossible to reason with him.'

Robinson did go to see Fry. The detective told him that he had no intention of dropping the case; that it was the clearest and strongest case he had ever brought to court. There was nothing more that Robinson, the great opponent of I.D.B., could do.

But Barney had no intention of giving up so easily. He decided to tackle Fry himself. His approach was casual. Meeting the detective in the street, he invited him to have a drink at the Barnato Bar. While they were drinking, mention was made of a 'fancy diamond pin' belonging to Fry. According to Fry, Barney spoke of the pin first and asked whether he could see it. Fry said the pin was in his office and offered to show it to Barney there. They went to the detective's office. 'I showed him the diamond pin and he admired it greatly,' said Fry. 'I asked him what he thought was the value of it, and he said, "If I were you, I would not take £1,000 for it." '

Barney, as the detective knew, was one of the shrewdest diamond merchants in Kimberley and was well aware how much the pin was worth. Fry also had a good idea of the value of the pin. It was a 2 carat cut diamond which could be expected to fetch £100 at the outside. (Fry actually sold the pin for that amount later.) The drift of the conversation was obvious. But Barney was too nervous to make an outright offer. He told Fry that he wanted to discuss the illicit diamond trade with him but did not wish to speak about it in the office. They might, he said, be overheard. He invited the detective to call upon him that evening. To this Fry agreed but, on thinking it over, he decided not to take up the invitation. 'Something occurred to arouse my suspicions,' he said. This, from Kimberley's leading detective, rings a little hollow.

However, a couple of days later Fry was passing the Barnato Building and Barney called him into his office. This time he came straight to the point. He told Fry that he was extremely upset about Isaac Joel's arrest and the disgrace the case would bring upon his family. He said that he was prepared to pay Fry £5,000 for his diamond pin on condition that the charge against his nephew was withdrawn and Joel's release obtained. The detective refused the offer and immediately reported the matter to the Crown Prosecutor. There was nothing more he could do. His interview with Barney had been conducted in private and, there being no witnesses, he was advised to take the matter no further.

But Barney persisted. On the morning of 24 May, he went to the detective's house. He was frantic. 'He seemed distressed,' said Fry. 'He said, "If you knew how ill and distressed the poor boy's mother is you would have mercy on him." He said that he himself was very much distressed. He then bared his arm, and showed me some eruptions, brought on, as he said, by distress.' Unlike Robinson, Fry was not impressed by Barney's boils. Nor did he take kindly to a further offer of £5,000 for his diamond pin. He advised Barney to consult the Crown Prosecutor.

Three days later the Joel case came up for hearing in the Special Court. Isaac was not there to face the charge. He had bolted. Whether he had been told to leave the country by his uncles is not certain. But he was never seen in Kimberley again. Harry Barnato's visits to the Diamond City also became less frequent from this time on. The Kimberley side of the business was left in the hands of Barney and Woolf Joel. The £2,000 bail put up by Woolf Joel was paid in full but the Crown only managed to recover £346 of the £2,000 pledged by Isaac Joel.

The affair did not end there. Barney Barnato's reputation as an easy-going, affable clown still won him many friends. But when he was crossed or humiliated his neurotic, vindictive streak showed itself. He could be, and often was, utterly ruthless. He now set out to get his revenge.

[3]

In the months following Isaac Joel's disappearance, a campaign was started in Kimberley to discredit the Detective Department. Certain provisions in the Diamond Trade Act came under attack; the methods employed by John Larkin Fry to obtain convictions were vehemently denounced. In September 1884, Fry was suspended from duty. His suspension was announced in the first number of a new newspaper, the *Diamond Times*, which was published on 1 October 1884.

This newspaper was nominally owned by young Woolf Joel, but was known to be financed by Barney Barnato. Its editor was that old Kimberley journalist (and former adversary of J. B. Robinson) R. W. Murray. Murray had left the diamond fields in 1881 but had recently returned and now threw his weight and considerable newspaper experience behind the campaign against the Detective Department. There can be little doubt that Murray entered the campaign convinced of its righteousness, but he undoubtedly contributed to the downfall of John Fry. Largely as a result of the campaign (and perhaps some backstairs influence in Cape Town) Fry was formally dismissed from the Kimberley police force in February 1885.

Barney Barnato was cock-a-hoop. All his old confidence returned and he went about boasting of his future plans. It was rumoured that he intended moving to Cape Town and was considering contesting a Cape seat in the House of Assembly. Unfortunately, his gloating went a little too far.

Shortly before John Fry's formal dismissal, Barney met the detective in the Queen's Hotel, Kimberley. He was geniality itself. He took Fry into one of the inner rooms and offered to shake his hand. When Fry asked why they should shake hands, Barney replied: 'I am a good friend, but a bitter enemy.' The detective, however, was in no mood for patching up old quarrels. As far as he was concerned the feud was still on. 'I suppose', he said, 'I have to thank that scurrilous print which you support for the position I am placed in at present?' According to Fry, Barney admitted that had Isaac Joel been let off, the *Diamond Times* would never have been started.

A few days later Fry, who had been on a short holiday, travelled back to Kimberley in the same cart as Francis Dormer, the editor of the *Cape Argus*. He told Dormer about his ill-fated dealings with the Barnato family. Dormer was immediately interested. As editor of the *Argus*, he was particularly concerned with affairs on the diamond fields. He had good reason to be. His employer was Cecil Rhodes. Although it was not generally known, Rhodes had advanced Dormer £6,000 to buy the newspaper in 1881. The take-over had been arranged with the utmost secrecy. An agent of Rhodes's had met Dormer on the Grand Parade in Cape Town and had handed over the first instalment of the purchase price—£3,000—to ensure that details of the transaction did not leak out. Rhodes wanted the *Argus* to support him and retain the semblance of an independent news-paper. Any opponent of Rhodes's was guaranteed to receive short shrift from Francis Dormer.

Dormer published his own version of Fry's story in the *Argus* on 27 December 1884. He got his facts rather muddled. After asking for Fry to be given a fair trial and claiming that he had been victimised by 'a certain resident of Kimberley not unknown to fame', he went on: 'If the jade Rumour does not lie, this gentleman, one of the Barnato Brothers—"Barney" of that ilk, told Mr Fry he would not have lost his situation if he had accepted the £5,000 offered him for a worthless diamond pin which he wore when the proceedings were commenced against one Wolf Joel.' He then explained how the Joel diamond registers had been impounded and how 'Wolf' Joel had escaped trial. 'The bail was estreated,' he wrote, 'and Barnato, who was security in reality though not in name, cheerfully paid the £2,000, glad, even at that price that his kinsman for whom he had entertained no small amount of genuine affection, should escape the meshes of the law. The Crown authorities at the Fields were of the opinion that the impounded books disclosed a partnership between Barnato and Joel and they were anxious to proceed against the former and try, at any rate, to make him responsible for the remain-ing £2,000, but leave from Cape Town was not obtained.'

The day after this was published, Dormer was told that he had made a mistake in naming Woolf Joel instead of Isaac Joel. He immediately published an apology. It was too late. Both Barney and Woolf Joel sued for libel. The case caused a tremendous stir in Cape Town and Kimberley. Among other things it was rumoured that the *Argus* was owned by Cecil Rhodes and the libel action was part of a financial vendetta. Dormer quickly denied that there was any connection between the *Argus* and Rhodes. One wonders how far Rhodes *was* implicated in this attack on Barnato; it might well have been the first of their many clashes.

In the action brought by Woolf Joel, Dormer had little defence. He said that he had made a genuine mistake and gave his published apology as evidence. The jury found him guilty, with a recommendation for leniency. He was fined £10.

He had very little difficulty in justifying his remarks about Barney. It was largely a matter of Barney's word against Fry's. The only other important witness called was J. B. Robinson, whose testimony tended to support that given by Fry. Barney himself put up a poor show. He flatly denied everything the detective said and claimed that he was sorry that Isaac had escaped trial. His case was not helped when his own counsel refused to pursue the innuendo that his business methods were dishonest. The jury found Dormer 'Not Guilty' of publishing a defamatory libel on Barnato.

The 'Argus Libel Case' was followed eagerly throughout South Africa. Reports of the court proceedings were wired direct to Kimberley and it was estimated that over 45,000 words had been cabled 'keeping the Telegraph Office open until 2.30 a.m.'. The verdict met with loud approval. Even R. W. Murray, in the *Diamond Times*, had to admit that: 'There is no doubt whatever that the verdict in the case of Barnato vs Dormer is a victory for the Colonial Press.' Neither Barney nor Woolf Joel received much sympathy at the diamond diggings. 'The public feeling on the subject of the "Argus" libel case', said the *Diamond Fields Advertiser*, 'is shown by the fact that, at Dutoitspan, and in the course of an hour or so, the ten pounds fine inflicted on Mr Dormer in Woolf Joel's case was made up in half-crown subscriptions. We hear of ready contributions by the leading citizens of Kimberley, De Beers and Bulfontein.'

Barney's popularity was by no means what it had once been.

[4]

None of Barney's biographers mention this unhappy episode. By ignoring it (or not knowing it) they fail to appreciate a fundamental defect in his character. Barney is mostly portrayed as he wished to be portrayed; a rascal perhaps, but a lovable rascal, always good for a laugh, never taking himself too seriously. This is how a great many people saw him during his lifetime.

'As he, in his gleeful moment plays the jester today, with his merry quips and shrewd observations, so he played it yesterday, and so he will play it tomorrow,' says Stuart Cumberland, who knew him in later years; 'there has not been and will not be any change in this direction in Barnie [sic]; age will not sober him and money will not dull him.' But this is the surface Barney; Barney as he appeared when things were going his way. Inwardly he was plagued by a basic insecurity, a suppressed hysteria; emotional instability for which all the money in the world could not compensate. Once the clown's mask slipped his psychological failings were exposed. The Barney Barnato described by J. B. Robinson and John Fry was not merely a man troubled by adversity; he was a man who tortured himself beyond reason. It is necessary to understand this if one is to appreciate his ultimate fate.

There is another important omission in books about the Barnato family. The name of Isaac Joel is never mentioned. Isaac is always referred to as John (Jack) Barnato Joel. This is the name he adopted when he eventually returned to England. He was the last of the Joel brothers to control the Barnato fortunes; he died a multi-millionaire.

Only those with long memories could recall his unfortunate sojourn in South Africa. One of them published an anonymous attack on him in England at the beginning of the present century. At that time he was living, as J. B. Joel, at 34 Grosvenor Square in London on a fortune estimated at £4,000,000. He lived in great style. 'The door of your town mansion', it was said, 'is opened instantly in the most open-Sesame, up-to-date fashion, and men servants, who show evidence of their having been trained in our best old English families, stand stately and obsequious in the hall. . . . Lavishly furnished, gorgeously upholstered, the mansion is a receptacle for everything fashion can light on, and resembles a gilded pantechnicon.' This splendour was contrasted with Joel's earlier career. His attacker speculated on what happened after he fled Kimberley: 'Rumour says you decamped to Spain; at any rate, you shook the dust of the Cape Colony from your feet. Whether the alteration in the political atmosphere in British African possessions since the [Boer] war has altered the law as it previously existed we do not know, but as yet you have not revisited the country from which you have obtained your fabulous wealth. You might take for your motto, "Discretion is the better part of valour", which you so successfully enacted.'

The author of this attack was, it seems, Lou Cohen; but the man responsible for publishing it was Robert Sievier, the somewhat disreputable proprietor of the *Winning Post*. Sievier was also a race-horse owner and once, after losing a race to Joel, had blackmailed the millionaire by threatening to expose the secret of his South African past. Joel had paid the blackmailer several thousand pounds but had later changed his mind and had him arrested. In the trial that followed Sievier—brilliantly defended by Rufus Isaacs—had been found 'Not Guilty'. 'Mr Joel underwent a most damaging time at the hands of the redoubtable Rufus Isaacs,' it is said, 'half-way through the case he was heartily wishing himself out of it.'

'Jack' Joel, however, had been able to live down both this scandal and his youthful misdemeanour. By the time he died, in 1940, most of those with long memories had predeceased him. Indeed, he had taken his place among the heroes of South African finance. 'Men sometimes say unkind things about the Joels—Solly and Jack', wrote a columnist on the *Daily Mail*, after the first World War. 'But I shall always remember a remark I heard General Smuts make over a luncheon table at the Savoy. . . . "The two men who saved South Africa", he said, "were the brothers Joel." '

CHAPTER THIRTEEN

RHODES FINDS AN ALLY

By 1884, Cecil Rhodes was known in the Cape Assembly for more than his defence of the diamond industry. Interested, as he undoubtedly was, in Kimberley, he regarded the entire continent of Africa as his real sphere of activity. His dream of extending the British Empire and bringing peace to the world was no longer an undergraduate ambition, confided only to the wills he made in secret. He now spoke openly of his aims.

'All this to be painted red; that is my dream,' he would say, flattening a large hand on a map of central Africa. Not only did he mean it, he began to do something about it. He spoke often and long on African affairs. In July 1882, when a Commission was set up to investigate the claims of loyal Basutos, Rhodes was one of its members. He went to Basutoland and there met the man who was to become famous as General Gordon of Khartoum. Gordon had been impressed by the young Rhodes. 'Stay with me here and we will work together,' he had said. But Rhodes refused. He had his own ideas, his own plans. Gordon appreciated his obstinacy. 'There are very few men in the world', he said, 'to whom I would make such an offer, but of course you *will* have your way. I never met a man so strong in his own opinion; you think your views are always right.'

Others, nearer home, were no less impressed. In the House of Assembly his dogmatic pronouncements and his obvious enthusiasm had singled him out as a man to be watched. The doubts expressed after his maiden speech were soon dispelled. His speeches were deceptive. He never spoke well. He was far too excitable; he tended to ramble and had little control over his voice which often broke into a high falsetto. But those who listened to him closely quickly recognised the soundness of his arguments. He was logical, he gave the impression of sincerity and he could, at times, be inspiring. Before long he had developed a following among men who mattered.

Political debate was not really his *métier*. He was far too impatient. When not speaking himself, he became restless. His habit of sitting on his hands and bouncing up and down during a long-winded discussion caused others to become embarrassed. At times the seemingly pointless grind of parliamentary procedure drove him to despair. 'Politics to me are perfectly hopeless,' he wrote to J. X. Merriman in 1883. 'I shall stand again and believe I shall be returned but I have not much heart in the matter. . . . I am looking forward to being able to stroll round the world for a couple of years and intend doing so immediately after the next session.' But such moods passed. He was far more interested in changing the world than in strolling round it. By the end of the next session he was too busy to think of taking a holiday.

In March 1884 he was given a seat in the Cabinet as Treasurer of the Cape Colony. Great hopes were held out for his first Budget. 'You will see we have taken Rhodes into the Ministry,' Merriman wrote to a friend, 'he is a man of under thirty but of the greatest talent and originality, and I look upon him as by far the rising man in South Africa.' Merriman's wife, however, did not share her husband's enthusiasm. She considered Rhodes to be unstable, not able to manage his own financial affairs, and lacking in business acumen. Writing to Merriman, who was also in the Cabinet, she said: 'Well! All I can say—the fate of the Ministry is sealed, and I give you three weeks after Parliament meets. I am *thoroughly* disgusted. . . . The small interest I still had left in your Ministry has vanished with this new introduction, and for the sake of the country I can only hope you may be turned out soon.'

She got her wish. The Ministry fell a few weeks later. The Prime Minister's resignation had nothing to do with Rhodes, but it prevented him from presenting his Budget.

Rhodes's next appointment was more fortunate. It was also to his liking. In August 1884, he persuaded the British High Commissioner, Sir Hercules Robinson, to appoint him as a Deputy-Commissioner in Bechuanaland. He was about to take his first step north.

Bechuanaland was a vast and largely barren territory situated to the north of the Cape Colony and to the west of the Transvaal. Rhodes regarded it as essential to his plans for painting Africa red. In many respects it was an unpromising and unattractive country; its fascination for Rhodes lay in the comparatively fertile strip of land that ran along its eastern border—the border it shared with the Transvaal. Along this strip missionaries had passed on their journeys into the heart of Africa; it was known, in fact, as the Missionaries Road to the North. Where missionaries could go, so could trade, railways and, of course, British settlers. Rhodes's eyes had been turned to Bechuanaland from the moment he entered Parliament. 'I look upon this Bechuanaland territory', he told the Cape Assembly, 'as the Suez Canal of the trade of this country, the key of its road to the interior.' He was determined that Bechuanaland should be annexed to the Cape.

This was easier said than done. There were others who also had designs on Bechuanaland. The Boers of the Transvaal had already made advances into the territory. They had, in fact, set up two small independent republics—Stellaland and Goshen—at the mouth of Rhodes's 'Suez Canal'. If these republics were allowed to remain it would not be possible for British enterprise to flow northwards. In April 1884, the situation became further complicated when Bismarck proclaimed a German protectorate over the ill-defined country to the west of Bechuanaland. This awakened British fears. The scramble for Africa among the countries of Europe was under way and Britain recognised the need to secure an interest in Bechuanaland. A resident British Deputy-Commissioner, the Rev. John Mackenzie, was sent to the territory. Mackenzie, an ardent Imperialist and rabid anti-Boer, had caused so much trouble among the Boers of Stellaland and Goshen that he soon had to be recalled. This provided Rhodes with a long-looked for opportunity.

Rhodes had become totally disillusioned with British tactics in Africa. Whatever action the British Government took seemed invariably to lead to a reversal of British interests. He had no faith in the politics of a Government which operated from a distance of 6,000 miles and which appeared to take little cognisance of the opinions of colonial officials. The Bechuanaland problem could not be left to the vagaries of Whitehall. 'We want to get rid of the Imperial factor in this question and deal with it ourselves,' he declared. Mackenzie's recall had given him the chance to do just that.

As the clergyman's replacement, he set off immediately to win over the truculent republics. In Stellaland his down-to-earth diplomacy met with almost instantaneous success. 'Blood must flow,' growled Groot Adriaan De la Rey when Rhodes arrived. 'No, give me my breakfast, and then we can talk about blood,' replied

Rhodes. By the end of the week Rhodes had become godfather to
De la Rey's grandchild and had Stellaland in his pocket.

Goshen proved a trickier problem. The situation there was
complicated by tribal faction fighting. The Transvaal had already
proclaimed a protectorate over the territory and although, in response
to a British demand, they eventually withdrew their claim, Rhodes
still felt the need for a show of force. Unable to get support from the
Cape Government, he reluctantly fell back on the 'Imperial factor'.

In January 1885, at Rhodes's request, his old friend Sir Charles
Warren, with 4,000 troops, arrived in South Africa and proceeded
to Bechuanaland. Rhodes, dressed in a ragged coat, dirty white
flannels and an old pair of tennis shoes, joined the British troops
soon after their arrival. He was present at the meeting of Sir Charles
Warren and Paul Kruger, the President of the Transvaal. This
took place at Fourteen Streams on the Vaal River and was one of
the few times that Rhodes came face to face with the man who was
to be his greatest adversary. Rhodes took little part in the discussions
but he did not escape the eye of the wily old President. 'That young
man will cause me trouble', Kruger is said to have remarked, 'if he
does not leave politics alone and turn to something else.'

The final result of the Bechuanaland negotiations was not what
Rhodes had hoped for. He and Sir Charles Warren were soon at
loggerheads and the territory was lost to the Cape. Despite Rhodes's
protests, Bechuanaland was divided into a Crown Colony and a
British Protectorate. The area had been safeguarded from the
Transvaal and Germany, but Rhodes still had to contend with the
'Imperial factor'. He was not happy about this. The obstructionists
of Whitehall might well prove antagonistic to his future plans. He
needed a free hand. His money had started him on his political
career but it had not taken him far enough. He was still dependent
on politicians and Government officials. Not until he was powerful
enough to act on his own initiative could he hope to realise his great
vision.

And the only way he could achieve such power was through his
diamond interests. This was something he realised only too well.
He brooded over it. 'When I am in Kimberley,' he told a friend,
'and I have nothing much to do, I often go and sit on the edge of the
De Beers mine, and I look at the blue diamondiferous ground,
reaching from the surface, a thousand feet down the open workings
of the mine, and I reckon up the value of the diamonds in the 'blue'

and the power conferred by them. In fact, every foot of blue ground means so much power.'

The more blue ground he obtained, the more power he would wield. He needed to control not only the De Beers Mining Company, nor yet the De Beers Mine; he must take over all the diamond mines in Kimberley. The need to amalgamate the mines was recognised by every capitalist in Kimberley. For Rhodes it now became a double necessity. In his mind the future of the diamond industry had become wedded to the destinies of the British Empire.

[2]

But, although Rhodes resented the 'Imperial factor', it served his purpose for the time being. Once the road to the north was secure in British hands, he could turn his attention to local affairs.

His immediate political objective was the unification of the four states of southern Africa: the Cape, the Transvaal, the Orange Free State and Natal. These four states would provide the base for his territorial ambitions. His immediate financial objective was, of course, the unification of the four mines of Griqualand West: De Beers, Kimberley, Dutoitspan and Bulfontein. These four mines would provide him with the financial backing he needed. There was a great similarity between the problems he faced. He recognised this. 'I have always been an amalgamationist,' he told a gathering at De Beers. 'I look on amalgamation as the only thing that can give a really permanent character to our industry. I have also worked at it with a special interest apart from mining, for I have felt that the task of reconciling the various interests in our mine is a very similar one to that which must fall to the lot of some of our politicians in the future, viz., the union of the various States in this country, as they both require great patience, mutual concessions and ample consideration for local feeling.'

He had gone a long way towards achieving the unification of the De Beers Mine. One by one, the De Beers Mining Company had succeeded in buying out, or incorporating, most of the smaller companies in the mine. 'Amalgamation steadily progresses and I think that some day our scrip will be worth something,' Rhodes reported to Merriman in October 1883. The following year, two of the more important companies fell into the De Beers net. In March 1884 an amalagamation between De Beers and the Baxter Gully Company was announced; in April 1884 De Beers merged with the Independent Company. The ultimate fate of the De Beers Mine was fairly predictable.

A larger question mark hung over the neighbouring Kimberley Mine. Most people regarded this mine as the key mine on the diamond fields. Whoever controlled the Kimberley Mine, it was thought, would eventually dominate the diamond industry. Rhodes would have none of this. He considered, and continued to consider, that two shares in the De Beers Mine were worth three in the Kimberley Mine. Nevertheless, he could not ignore the undoubted importance of the Kimberley Mine. Nor could he ignore the struggle for supremacy in that mine. The outcome of that struggle was by no means as certain as it appeared to be at De Beers.

When Anthony Trollope had visited the diamond diggings in 1877, the most important concern in the Kimberley Mine was the Baring-Gould Company. This company had continued to expand. By the usual process of amalgamation it had grown, changed its name, and was now the powerful Central Company. With the advent of company mining, its position had been challenged by the French Company. This was the company launched by Jules Porges of Paris and represented in Kimberley by the cool, astute German financier, Julius Wernher. The French Company had also increased its holdings in the Kimberley Mine and acquired important interests in the other mines. The Central Company and the French Company were undoubtedly the most important of the rival factions in the Kimberley Mine.

They had, however, two close challengers: not so much in size as in the strategic positions they held. One of these was J. B. Robinson's Standard Company—the largest and most enduring of Robinson's mining concerns. The other was the Barnato Mining Company. In 1884 the Barnatos had purchased, in Woolf Joel's name, six claims adjoining their own for the enormous sum of £180,000. This had strengthened their position considerably and made them a force to be reckoned with in the Kimberley Mine.

However, Rhodes was not immediately concerned with the heads of any of these four firms; not with Francis Baring-Gould or Jules Porges; nor with J. B. Robinson or Barney Barnato. The man he saw as his potential rival was not in the public eye. He had played very little part in the civic or political life of Kimberley. Occasionally he would write a tart letter to the press on mining affairs; he was

known as a man of business and a diamond buyer of some importance, but he could hardly be regarded as a Kimberley personality,
let alone a celebrity. His name was Alfred Beit.

[3]

Sir Percy Fitzpatrick, the well-known South African mining personality, was once questioned about Alfred Beit. 'Fitzpatrick', said
a journalist friend, 'you knew him well. There was something
Christ-like about Alfred Beit.' Sir Percy agreed. He said that he had
often thought this but had dared not say so for fear of being scoffed
at by those who did not know Beit. His friend said: 'Why be afraid
to say it? *You* knew him, *they* did not.'

One can understand Sir Percy's reticence. To call a self-made
multi-millionaire Christ-like is to invite ridicule. Rich men are
seldom saints; self-made rich men are often the reverse. And to place
Alfred Beit so firmly among the angels was, in any case, to stretch
his virtues a little too far. But, if Beit did not exactly rate canonisa-
tion, he had earned himself a reputation almost as rare. He was a
gentle, self-effacing, likeable (to many people, lovable) plutocrat; an
exceptional being indeed to rise amid the dust, dirt and diamonds of
Kimberley.

Alfred Beit had come to South Africa in 1875. He was then
twenty-two, having been born in 1853—the same year as Rhodes.
To look at, he was singularly unimpressive. A dumpy young man,
whose head seemed too large for his body, he gave the impression—
with his bulbous brown eyes, receding chin and tiny moustache—of
being weak-willed and not over bright. He was painfully shy; a
mass of nervous mannerisms. He would tug at his collar, twist his
moustache and bite the corner of his handkerchief. With strangers
he could rarely relax. Embarrassed himself he would embarrass
others. Nothing could be done to put him at his ease; he had been
nervous as a child, he remained nervous until the day he died.

Before coming to South Africa he had shown very little promise,
either as a scholar or as a business man. 'I was one of the poor Beits
of Hamburg,' he once explained; 'my father found it difficult even
to pay for my schooling, and you know that is cheap enough in
Germany; I had to leave before I had gone through *Real-schule*.'

The fault was not entirely his father's; poor the Beits might have
been, but they were not as poor as all that. His father, Siegfried

Beit, was a descendant of Portuguese Jews who had settled in northern Germany in the sixteenth century. The Beit family had established themselves as refiners of gold and silver but, in 1846, they had extended their business to deal in base metals and chemicals. Siegfried Beit had commenced his career in the family business but had later left to start a silk importing firm of his own. This was not the only break he made from family tradition. A year after his marriage to Laura Hahn—whose background was similar to his own—he and his wife had been converted to the Christian faith. Their decision to become Lutherans was the result not, it is said, of religious conviction but of the ghastly anti-semitism which would have threatened the prospects of their children in Germany. Alfred was the second of the Beits' six children; his father had taken him away from school early, not so much because he could not afford to keep him there, but because it seemed a waste of money to continue his education. His brother Theodore, who later became a Professor of Music, was the bright hope of the family. Alfred, it was thought, might do better earning his own living.

A story is told of Alfred Beit's schooldays which, in a way, characterises the child he was and the man he became. He went to the private school of a Dr Schleiden at Hamburg. The reports he brought home all told the same story: his behaviour was good, his work poor. His parents became worried. Then one day he arrived home full of joy; he had been moved, he said, to the top of the class. Naturally his mother and father were delighted; they saw the beginnings of a successful academic career. They were soon to be disillusioned. A few days later the headmaster called and told them that Alfred's success existed only in Alfred's imagination; his work had not improved. Shamefacedly Alfred owned up. He had wanted, he said, to make his parents happy for a few days at least.

The story has a ring of truth about it. Much of Alfred Beit's happiness came to him through the happiness of others. This is particularly true where his mother was concerned: he adored her. As a boy his only known ambition was to earn enough to buy his mother a carriage and pair. He did this on his first visit home from South Africa—throwing in a coachman for good measure. His mother fully returned his devotion. She was the one member of the family who had faith in him; the bond between them was very strong.

On leaving school, Alfred was apprenticed as a clerk to a diamond dealer in Amsterdam. His family was connected with the Lipperts of

Hamburg—a firm of gem importers—who promised him an opening once he had learned the diamond trade. He was in Amsterdam for two years and learnt a good deal about diamonds, but he does not appear to have distinguished himself. 'I just did my work and wasted my spare time like other young men,' he was to say.

In 1875, the Lipperts, who had branches in Cape Town and Port Elizabeth, sent him to South Africa. He went first to Port Elizabeth but was transferred, almost immediately, to Kimberley.

The year 1875 was a bleak year on the diamond fields. It was the year of the Black Flag Rebellion and trade was in a depressed state. The loss of confidence among the diggers and diamond buyers was reflected in their attitude towards the diamonds being mined in Kimberley. A slump in the market had convinced many of them that South African diamonds could not compete with those sold from the long-established mines of Brazil. Beit knew enough about diamonds to know how mistaken this attitude was. He also recognised the opportunities it presented. 'When I reached Kimberley,' he said, 'I found that very few people knew anything about diamonds; they bought and sold at haphazard, and a great many of them really believed that the Cape diamonds were of an inferior quality. Of course, I saw at once that some of the Cape stones were as good as any in the world, and I saw, too, that the buyers protected themselves against their own ignorance by offering generally one-tenth part of what each stone was worth in Europe. It was plain that if one had a little money there was a fortune to be made.'

By offering high prices for diamonds he was able to attract customers. Soon he had built up sufficient good-will to cut his connections with the Lippert firm. With the money he had made, and an additional couple of thousand pounds raised by his father, he went into business for himself. His first investment, however, was not in diamonds but in property. Kimberley was growing fast and there was an urgent demand for houses and offices. Once again Beit seized his opportunity. He bought a piece of land and on it he built twelve or thirteen offices, keeping one for himself. They were little more than shanties but they brought him a monthly rental of £1,800. He held on to the land for several years, selling it eventually for £260,000. 'I got something for the dwellings, too, I think,' he was to say. 'Not a bad speculation.'

[4]

Beit's property investment launched him on his career. Hopeless at school, he had an exceptional mind when it came to business. His approach to finance was largely instinctive, but uncannily accurate. He was helped by an extraordinary facility at mental arithmetic. It was said that his mind was like a ready reckoner: at a glance he could reduce the most complicated balance sheet to simple terms. But it was his remarkable intuitive powers which gave him his touch of genius. Without making the slightest effort he seemed to know precisely when and how to act; many of his decisions appeared impulsive but they were rarely mistaken. So unfailingly did his hunches pay off that there were times when it looked as if he was actually making money against his will. Certainly he displayed none of the power-lust of other financiers; he was singularly unambitious; he did not seek success, it came to him.

Once he had built his shanties, he settled into the one he had selected for himself and commenced business as a diamond buyer. He did not have to look for sellers. His reputation for honest dealing soon spread. Not only was he prepared to pay good prices for diamonds but he was always ready to help anyone down on his luck. Diggers trusted him and he trusted them. Often he acted as a money changer. There was always a chronic shortage of small change at the diamond diggings and, at times, diggers were hard put to find silver with which to pay their African labourers. Every Saturday Beit kept a large bag of silver on his counter and diggers looking for change were told to help themselves from it. This so impressed the diggers, it is said, that none of them would have dreamt of taking a penny more than the exact change.

Not only was he honest, he was extraordinarily kind hearted. His generosity became a Kimberley legend. J. B. Taylor, who worked for him, tells how one day a tearful girl came to Beit's office with a note from her mother. After reading the note and asking a few searching questions, Beit wrote out a cheque and gave it to the child. When Taylor teased him for being so trustful, Beit let him read the note. It was from a colonial woman who had lost her husband and had a large family to rear: she had asked for money to purchase a small shop. On the slight evidence of the note and the little girl's answers, Beit had handed over a cheque for £250.

Being shy and nervous, Alfred Beit was invariably attracted to men of assurance and cool judgement. Such a man was Julius Wernher—the impressive-looking partner of Jules Porges. In 1880 Beit joined forces with the Porges concern and his working partnership with Wernher proved one of the most successful in Kimberley. They appeared an ill-assorted pair—the plain, diffident little Beit and the handsome, commanding Wernher—but as partners they were well matched. Lionel Phillips, who once worked for J. B. Robinson and later worked for Wernher and Beit, has noted the complementary facets of their characters. 'Of all the men I had become acquainted with at Kimberley,' says Phillips, 'none was more genial and kind, none more brilliant in capacity, more bold in enterprise, or more genuinely respected and admired than Alfred Beit. His intelligence was keen and his power of decision great as it was rapid. He and his partner, Julius Wernher, were, as business men, a unique combination. Beit had the gift of quite unusual insight, coupled with boldness of action, while Wernher had a calmer, colder and safer judgement.' They prospered exceedingly.

Beit early recognised the necessity for amalgamation. Like many others, he regarded the Kimberley Mine as the most important factor in any unification plan. Therefore his initial efforts were directed towards acquiring large interests in that mine. He became a shareholder of most of the important Kimberley Mine companies, concentrating, in particular, on the working of the Central Company. The Central Company had, in fact, achieved its dominant position largely through the skilful manoeuvring of Alfred Beit. 'At first', says one of the Company's directors, 'he worked through the Kimberley Central board but always keeping himself in the background. Under his quiet influence, not only with the Central but with other boards, the work of company absorption by our Central Co. went on until "The Kimberley Central" was practically Kimberley Mine.' But not everybody shared Beit's bold spirit. Some of the other directors, including the head of the Company, Francis Baring-Gould, took fright. The diamond industry was far from flourishing and they were wary of taking on further responsibilities. Beit was balked.

However, his labours had not passed unnoticed. At De Beers, Rhodes had watched the growth of the Central Company with no small interest. He recognised Alfred Beit as a man to be taken seriously. And, for Rhodes, to take someone seriously meant coming to terms with them; or, as he was fond of saying, winning them over 'on the personal'. Nobody was more completely won over to Rhodes 'on the personal' than was Alfred Beit.

[5]

It is not certain when Rhodes and Beit first met. W. P. Taylor was to claim that he introduced them to each other in 1879. But it would appear that Taylor got his dates mixed. Beit himself implies that he and Rhodes did not get to know each other until after he had entered into partnership with Julius Wernher in 1880. This is borne out by Beit's activities. Throughout the early 1880s, he was busy building up the Central Company and was hostile to De Beers. He was outspoken in his hostility. In March 1884, for instance, when an enquiry was held into the financial state of the Mining Boards, Beit led a delegation protesting against the undue influence of Rhodes's company at the De Beers Mine. He complained that four of the seven members on the De Beers Mining Board were sitting for one company—De Beers Mining Company—and that the smallholders were suffering in consequence. It may have been then that Rhodes decided to win Beit over. The following month Beit attended a meeting of the De Beers Mining Company as a shareholder.

A story is told of the coming together of Cecil Rhodes and Alfred Beit. There could be some truth in it. One evening Beit is said to have been working late in his office; he was puzzling over his amalgamation plans. Rhodes looked in on him.

'Hullo,' said Rhodes, 'do you never take a rest?'

'Not often,' answered Beit.

'Well, what is your game?' asked Rhodes.

'I am going to control the whole diamond output before I am much older,' said Beit.

'That's funny,' said Rhodes, 'I have made up my mind to do the same; we had better join hands.'

And join hands they did.

The became business partners and they became friends. Rhodes had become slightly more sociable since sharing a cottage with Neville Pickering; now he was often seen about Kimberley with Alfred Beit. They drank champagne and stout together at the Craven. They played poker—badly—at the Kimberley Club. Occasionally they were to be seen at a Bachelor's Ball: Rhodes vigorously twirling the plainest girl in the room and Beit prancing beside the tallest. The diminutive Beit had a *penchant* for tall ladies. To watch him dancing was one of the sights of Kimberley. He appeared, it is said, to run round his lofty partners rather than dance with them. In Rhodes, Beit found another dynamic and purposeful hero. From the time of their meeting, Beit devoted his life to Rhodes's interests: nobody surpassed 'little Alfred' in his loyalty to Cecil Rhodes.

Rhodes, in turn, came to depend upon Beit's financial advice. Those who knew them both refer to Beit as Rhodes's 'financial genius'. Any problem concerning diamonds would invariably be solved by Beit. 'Ask little Alfred', became a catch phrase among Rhodes's friends. Together they made a formidable team. When they undertook to monopolise the mines, they set the pace of the race towards amalgamation. Nevertheless, they were given a good run for their money.

CHAPTER FOURTEEN

CLEARING THE FIELD

BY the 1880s Kimberley had developed from a jumble of tents into an untidy, unplanned and unattractive town. Wedged between two great holes—the Kimberley and the De Beers mines—its narrow streets followed no pattern, its buildings claimed little architectural merit, its open spaces boasted few trees. It was crowded, close-packed and all higgledy-piggledy. With bricks still in shorter supply than diamonds, the town was built, almost entirely, of corrugated iron. Complicated wooden trellis-work would enhance the corrugated iron walls of some of the wealthier homes; a one-storey brick façade —pompously pedimented and pilastered—would screen a mere corrugated iron shed. The very flower pots lined up on the verandahs were fashioned from old meat tins. Hot, dusty and haphazard, Kimberley was still little more than a shanty town; a conglomeration of noisy bars, ill-ventilated shops, one-room mining offices and ramshackle hotels.

Yet there were indications that it harboured some of the wealthiest men in South Africa. During the 1880s an impressive Club House, two-storeyed, red bricked and wide verandahed, was built in Dutoitspan Road. Here and there the turrets of ornate private homes rose above the carefully nurtured blue-gum trees. Richly patterned cast-iron pillars and balustrades graced an occasional public or private building. One might catch a glimpse of a superb carriage, a plush interior, even a uniformed servant.

It was for its vitality, however, that Kimberley was chiefly remarkable. What it lacked in attractions, it more than made up for in dash. It had retained much of its mining camp atmosphere. In fact, the older inhabitants still referred to the areas surrounding the mines as camps. De Beers was merging into Kimberley and Dutoitspan was now the municipality of Beaconsfield, but the old localities were remembered and old rivalries remained. Lively and competitive, the town took pride in its ceaseless activity.

'The streets of Kimberley are never at any time particularly dull, as compared with those of most colonial towns,' boasted a citizen in 1885. 'There is always to be heard the crack of the Cape driver's whip, as he rattles his fare off to the sister Camp of Beaconsfield, or the 'Pan' as old residents still delight in calling it. From early morning until the setting of the sun, bullock wagons, with their spans of from 12 to 18 oxen, continuously rumble along, to and from the various Camps, coming with their heavy freight of wood and produce, or returning with lighter loads and heavier pockets, figuratively speaking. The constant hurrying to and fro in the vicinity of the Market-square of produce merchants, Boer farmers, horse dealers, auctioneers and their assistants, storekeepers, lawyers' clerks, miners, Kaffirs, Malays, Indians, and the occasional and irrepressible Celestial, constitute a conglomeration of busybodies which may, perhaps, imbue the casual visitor from some dull country town with the idea that, although a considerable item in connection with Kimberley life may consist of "beer and skittles", so to speak, real downright hard work also forms a prominent portion of the day's doings.'

The casual visitor might, indeed, have been impressed by Kimberley's industry. He may well have thought that the bustle in the streets reflected booming prosperity. But he would have been wrong. The economic crises which plagued the diamond industry throughout the eighties were far more serious than those which had driven the diggers to rebellion in the mid-seventies. If they lacked the drama of widespread revolt, they often produced violence and were every bit as crippling. More so. At times it looked as if the industry would collapse completely. It took a long time before the real cause of the trouble was recognised; or, at least, before it was admitted. All too often those seeking a remedy were diverted by side issues which were only partly relevant to the basic situation.

The first serious blow fell towards the end of 1882. A series of disastrous land falls put many companies out of business. The Kimberley Mine was the worst affected. Among the companies whose claims were buried in this mine were J. B. Robinson's Standard Company and the Barnato Mining Company. Both these companies were forced to stop work. Of the other companies, a few of the richer concerns were able to carry on with greatly increased costs; many smallholders were ruined.

The ensuing crisis revealed a fundamental weakness in Kimberley's economy. Most of the companies had been financed by local capital. The hopes of enticing foreign investors had not been realised. The loans now demanded by the smaller companies to meet their increased costs could not be met; there was insufficient capital available. The poorer diggers were forced to sell up and the value of shares in the larger companies plummeted. A long period of depression set in.

'The decline in the value of shares in the market was enormous', says Dr Mathews. 'Central shares in the Kimberley mine, which had an easy sale in March, 1881, at £400 per share were in 1884 almost unsaleable at £25. Rose-Innes shares which were sought after at £53 sank to £5, and a similar fall also occurred in the shares of all the companies in the other mines of the province.'

Nevertheless, the firms which were soundly financed were able to profit from the slump. At this time, when Central shares were so low, Alfred Beit was able to extend the Company's holdings.

The situation was confused. If there was a shortage of capital, it was thought to be due to the fact that the diamond output was not realising its true value. One of the reasons for this—and it was thought to be the most important one—was that traffic in illicit diamond buying had deprived the industry of its just monetary return. The almost hysterical support given to Robinson and Rhodes during the passing of the Diamond Trade Act, came from those who regarded I.D.B. as the source of all Kimberley's evils. The crime was made to account for all the industry's misfortunes.

'It is the fashion to ascribe the failure of many companies to work profitably claims which have always yielded a handsome return in the days of individual diggers to the depredations of the I.D.B.,' the *Independent* had said in 1881; 'and, no doubt, these scoundrels are responsible for a large portion of the want of success. But they are not the only causes of failure. There is much mismanagement on the part of the companies themselves.'

Such strictures went unheeded. The stamping out of I.D.B. was looked upon as the first step towards prosperity. But it had not happened. The passing of the Diamond Trade Act coincided with the start of the depression. This had led mine owners along another dangerous path.

The Act, in itself, was thought to be insufficient to combat I.D.B. Stricter measures would have to be taken in the mines. The result was what had been prophesied: the African labourers were housed and confined to compounds and a system of searching all workers was instituted. Searching huts were set up at the entrances of the mines and every employee, white or black, was expected to submit to a close examination before leaving work. White overseers, who had previously ignored warnings that this would happen, now rebelled. A mass meeting was held on the Kimberley race-course and angry speeches were made. 'Who were to search the searchers?' became the cry. A resolution was passed protesting against 'the obnoxious searching system' and calling for overseers to strike the following morning. The next day no white men went to work.

The strike continued sporadically for months. A climax was reached in April 1884 when a large body of mine workers rioted. They marched on the mines, in an attempt to stop strike-breakers from working, and were fired upon by special police. Four of the strikers were killed outright and two of the many wounded died later.

This did not end the strikes. Eventually the mine owners were forced to modify the searching system, as it applied to white men, and the overseers returned reluctantly to work. They had merely succeeded in worsening the economic situation. The real problem remained unsolved.

It had nothing—or very little—to do with I.D.B. The fault lay not in the stones that were stolen but in the stones that were marketed. The diamond itself was the enemy of the industry. For most diamonds are useless. Their value is artificial and depends largely upon their scarcity. Before the opening of the South African mines they *were* relatively scarce. By flooding the market with stones, the Kimberley diggers had cheapened the product of their labours. The price of diamonds tended to fluctuate; more often than not to the seller's disadvantage. This danger had long been appreciated by the wiser diggers. Men working at New Rush, as far back as 1871, were said to blanch when they heard rumours of new discoveries. The last thing any of them wanted was for others to unearth an excess of diamonds. That was the trouble. Only if some stopped selling would the others benefit. But who could be expected to stop selling? They preferred to find other excuses for their ills.

Cecil Rhodes is said to have approached the problem with characteristic shrewdness. He assumed that every young man who became engaged would want diamonds for his future bride. The number of stones that these young men bought depended entirely on the state of the diamond market: if diamonds were cheap they would buy more, if expensive, fewer. But the amount they were prepared to spend on diamonds would not vary much. Taking into account the number of engagements that could be expected in a year, he estimated the annual demand for diamonds at four million pounds. This, he felt, was a rough but reasonable target for diamond producers to aim at. However, his calculations did not, as has been supposed, lead him to advocate that the diamond output should be regulated. For a long time he was opposed to such a suggestion.

It was J. B. Robinson who first tackled the problem publicly. At the beginning of July 1883, he called a meeting to discuss the possibility of regulating the production of diamonds. Opening the discussion he spoke at great length about 'the ruinously low prices induced by the enormous overproduction'. The only solution was, he said, to restrict the output of the mines by common agreement. Rhodes disagreed. He took the view that it was first necessary to fix a price below which diamonds could not be sold. They must agree among themselves, he said, not to sell for less than the cost of production. As so often happened when Robinson and Rhodes met, the discussion developed into a heated argument. Finally they agreed to hand the matter over to a committee formed from representatives of the four mines. But nothing more seems to have been done about it.

There was, in any case, little hope of either plan working. To attempt to regulate the price or the output of diamonds on a voluntary basis was to try the impossible. Competition was too keen and corruption too rife for a gentleman's agreement to work in Kimberley. Any form of regulation would have to be enforced; this could only be done by a company or a combination of companies which controlled the entire industry; no dissension could be allowed. The answer was amalgamation. The question was: how and by whom would amalgamation be achieved?

[2]

The meeting called by Robinson was held in the middle of 1883—the year when things seemed at their worst in Kimberley. It was the year in which the 'great strike' started and a year which ended in further disaster. Towards the end of 1883, a small-pox epidemic broke out in Kimberley. Alarming as the spread of the disease was, the controversy it aroused was scandalous.

Small-pox epidemics were by no means unknown in South Africa. They were, in fact, an ever-present threat. In 1882 a serious outbreak had occurred in the Cape Peninsula and only the prompt action of the mining community—led it is said by Cecil Rhodes—had prevented it invading Kimberley. By establishing a quarantine camp on the main road from Cape Town to Kimberley, supervised by a young Afrikaans doctor, Hans Sauer, who insisted that travellers to the town be vaccinated, the Kimberley authorities had managed to check the spread of the disease.

But a year later a gang of African labourers from Portuguese East Africa were suspected of having brought the disease to the diamond fields. They were immediately isolated and examined by a team of local doctors. These doctors reported that the disease was not small-pox but 'a bulbous disease of the skin allied to pemphigus'. The report was signed by a majority of Kimberley's medical practitioners, including Dr Mathews, Dr Murphy and the town's popular physician, Dr Leander Starr Jameson. Their findings, however, were challenged by Dr Hans Sauer who, from his previous experience, diagnosed small-pox. He claimed that the outbreak was being hushed up for fear that the African mine-workers would desert the town if there was a threat of an epidemic. This produced loud protests from the mining industry. 'It was not only the diamond magnates who were hostile,' says Sauer, 'but the vast majority of the population, who were in the same galley as the magnates, for should the mines shut down on account of this disease, they would starve just as certainly as the magnates would lose their profits.'

The campaign waged by Sauer, to refute the findings of his medical colleagues, resulted in bitterness on both sides. Several legal actions were instituted before it was established that the disease had spread to the mine-compounds and even into the Kimberley Hospital. Sauer's diagnosis was correct but proof came too late. The epidemic raged for two years, there were 2,300 cases and 700 deaths; of these 400 cases and 51 deaths occurred among the white population. J. B. Robinson played a prominent part in advocating measures to fight the epidemic, but he did not help matters by blaming the spread of the disease on the lack of control which Rhodes and the De Beers Mining Company exercised over their African workers. Disunity among the mining companies was as apparent in this crisis as in everything else.

The year 1884 saw no relief in the depressed state of Kimberley's financial affairs. 'Trade on the Fields continues very stagnant', became a standard summary of every market report. Almost as regular were the news items reporting cases of suicide among prominent citizens of the town. Hardly a week passed without an announcement of some well-known businessman taking his own life. Clergymen and rabbis lectured their congregations on the crime of self-murder and newspaper editors, weary of the sameness of their reports, lost all sympathy for the victims. 'The fact is that this mania for suicide taints the whole moral atmosphere,' declared the *Diamond Times*, 'and it is questionable whether it is good for the safety and morality of Society that such acts be recorded with tenderness. . . . There is never a hero without a struggle and a battle, and the battle of life is worth fighting, though a man may sometimes have to fight against desperate odds.'

Everyone was fighting against desperate odds. The Mining Boards were among the many institutions which collapsed. Unable to meet the demands made upon them or to collect dues from their members, they went bankrupt. An inquiry was held and in January 1885, J. X. Merriman was sent to Kimberley as a mediator between the Mining Boards and the Cape banks. Merriman's conclusion was that of every informed person: the crisis in the industry could only be solved by amalgamation. He was already working actively to bring this about.

It is incredible that so many of Cecil Rhodes's admirers firmly believe that he alone conceived the idea of amalgamating the diamond mines. They sometimes speculated on how far he was helped by Alfred Beit, but they are convinced that the idea originated with Rhodes. This, of course, is nonsense.

The idea of amalgamation was almost as old as the diamond diggings themselves. In 1872, Frederick Boyle, who had visited the diggings the year before, drew attention to the twin problems facing the industry. 'You cannot drown the market with an article only appertaining to the highest luxury,' he wrote, '—you cannot popularise a traffic in such articles—without swift and sudden catastrophe. These things require the most delicate manipulation, they exact the strictest reticence, they need a hand to hold them back or loose them as the occasion asks. . . . By royal monopoly alone, or by means of great and powerful companies, can jewel digging be made a thriving industry. Into the hands of a company all these public fields must fall, and, thus used, they may benefit the country for generations to come.'

Several attempts at amalgamation had already been made, of course. In 1875, at the time of the Black Flag Rebellion, the diggers in the Kimberley Mine had initiated a scheme of their own. They pooled the bulk of their claims and appointed a representative to offer the mine for sale in London. But the English lawyers who examined the scheme, found that such a sale was impossible. At that time the diggers held nothing more than a monthly 'lease-at-will' on their claims and were still restricted by the two-claim ruling. These negotiations had resulted in the Government giving the diggers a measure of fixity to 'claim-titles' and to the gradual removal of the claim restrictions. As soon as these restrictions were removed the diggers had tried again, only to be thwarted by J. B. Robinson who had his own scheme for amalgamation. In 1881, Rhodes had interested Baron Erlanger in his plan for amalgamating the De Beers Mine but this had fallen through. The following year Rothschilds of London had taken a hand in the game. They had sent a Mr Gansl to Kimberley to investigate the possibility of amalgamation, starting with the Dutoitspan Mine. Gansl had been enthusiastically championed by J. B. Robinson, but had been no more successful than the others. To weld the dissident elements of Griqualand West into a prosperous whole was a thankless and frustrating task. Even with such a glittering prize at stake, few had the heart to pursue it.

J. X. Merriman had been toying with the idea of amalgamation for some time. He regarded the stability of the diamond industry as essential to the well-being of the Cape's economy. After a visit to London in 1883, he had contacted J. B. Robinson and asked him what the chances were of amalgamating the Kimberley Mine. Robinson had been most cooperative. 'During your absence from the Colony,' he replied on 5 January 1884, 'I have worked very hard to bring about amalgamation of the mine on some fair basis, but unfortunately there are so many complications and conflicting interests at work that it seems impossible to carry out the measure no matter what efforts are put forth in that direction. It would be an excellent plan if you could act as arbiter between the various holdings in the Mine but it requires very careful management. . . .' He went on to suggest that Merriman should work through the Standard Bank in Cape Town. If the Cape Town branch of the bank approached the Kimberley branch, he said, it would appear that the scheme had come through an impartial channel and it would then be more acceptable to the general body of shareholders. 'I shall of course,' he assured Merriman, 'render every assistance. . . .'

This scheme had not proved feasible. However, Merriman had managed to get the backing he needed. In 1884 he had entered into discussions with C. J. Posno, a member of the well-known South American family, long connected with the Brazilian diamond mines. Posno had important interests in the mines of Griqualand West. The discussions were fruitful. In August 1885, Posno sent Merriman details of a syndicate which had been formed in London. This was the Unified Diamond Mines Ltd, established with a share capital of £10,000, for the purpose of buying as many properties as possible in the four mines of Griqualand West. Those companies which sold out to the Unified Company were to receive a corresponding proportion of shares. Two French banks had backed the enterprise with a guarantee of £600,000. Merriman was appointed the company's South African representative; A. Moulle was nominated as the overseas representative.

Merriman hailed the plan as a 'message of salvation' for the diamond industry. In January 1886, he and Moulle went to Kimberley to start negotiations. The first person they contacted was J. B. Robinson. 'Personally I am very anxious to see you take a leading share in the matter,' Merriman wrote to Robinson, 'and I was glad to find that Moulle has the same opinion of your abilities and influence. . . I am sure that your large interests and the faith you have always held in the future of the industry will give you a very important share in the next stage of its development.'

Merriman was in for a shock. A few days later he met Robinson to discuss the scheme. He discovered that the 'Buccaneer's' faith in the future of the industry was by no means what it had once been.

[3]

The depression had hit J. B. Robinson badly. He had been one of those to suffer from the beginning. The disastrous land falls of 1882 had put paid to the working of his most important concern, the Standard Company. Situated in the best part of the Kimberley Mine, it remained idle throughout the depression period. His other companies—such as the Rose-Innes Company—had not fared much better. The value of his shares had declined to such an extent that he had been forced to relinquish them. The only companies in which he retained large holdings were the moribund Standard, and the badly mis-managed Griqualand West Company in the Dutoitspan Mine. In theory these were two of the most important companies on the diamond fields; in fact they yielded next to nothing.

Only with substantial support from the Cape banks had Robinson managed to survive. In one way he had been fortunate. His name and influential position had enabled him to obtain loans; others had failed to do so and had been driven beyond despair. But the loans were enormous and his income had been reduced to practically nothing. The eighties had seen Robinson battling for his existence.

The long financial struggle had forced him to resign his seat in the Cape Assembly. On nomination day for the 1884 elections, Dr Murphy, not seeing Robinson at the nomination court, had taken it upon himself to put his friend's name forward. The next day, however, it was announced that Robinson had had no intention of standing and had notified the Civil Commissioner accordingly. Later Robinson was to claim that he had resigned his seat because his increasing deafness had made it impossible for him to take part in the debates. But this was not the real reason. His deafness did not prevent him standing for election two years later; he was quite prepared then to take part in debates. No: it was his precarious financial position which forced him to devote all his time to his own concerns. He did not like to admit that he was on the verge of bankruptcy: indeed, he never did admit it.

His efforts paid off. With the help of the managing director of the Standard Company, Mr Francis, he was able to get the company working again. For two years their valuable claims had been buried under debris in the centre of the Kimberley Mine; now they sank shafts and commenced underground mining. By May 1885 the company was once again able to declare a dividend: it was a dividend of only $1\frac{1}{2}$ per cent, but it was a beginning. Better times were prophesied. 'They must be proud of their success', said a financial columnist, 'for no Company has a sounder position than the "Standard". Reef may now fall to any extent for all the Company need care. The underground works are in famous order, and operations are being carried on as briskly as they can possibly be.'

But they could not be brisk enough to save J. B. Robinson. Success in the mine had come too late. Even if the declared dividend had been much larger it would have made no difference. He was the company's largest shareholder but his shares did not benefit him in the slightest. He had pledged them all against the loans he had received from the bank. And his difficulties had turned to disaster when the bank, insisting on the repayment of its loan, had begun to sell off the shares.

This was the position at the beginning of 1886, when Merriman arrived to ask his help in the new amalgamation scheme. To all appearances, Robinson was still the most important and influential of the Kimberley diamond magnates; in fact, he was standing helpless as his fortune was sold off by the bank. Unless he could obtain a further loan to stop the rot, he would be ruined.

Merriman was shocked. He had counted on Robinson's assistance. There was little he could do, but he did what he could. He gave Robinson a letter of introduction to Lewis Michell, the General Manager of the Standard Bank in Cape Town. The Standard Bank was interested in the amalgamation of the mines; Merriman thought that Michell might help. 'Mr Robinson,' wrote Merriman, 'is virtually, the Standard Co and the Griqualand West Co and apart from this he is beyond question the most influential person in Kimberley in a matter like the one we have in hand. His active cooperation would mean a very great deal more than the mere adherence of the companies and would probably secure the success of the project. I will endeavour to set before you the present state of Mr Robinson's affairs as he disclosed them to me. . . . Mr Robinson became liable to the Cape of Good Hope Bank for the sum of Forty Seven Thousand pounds (£47,000). . . . By way of security he pledged 600 shares of £100 of the Standard Co of the nominal value of £60,000 and 400 Griqualand West shares of the nominal value of £40,000. By means of dividends and other payments this debt has been reduced to somewhat about £28,000. The Bank has attained a complete cession of Robinson's shares and in face of the vital importance of amalgamation, it is now selling off these shares in small parcels to persons who will probably oppose the scheme at the same time sacrificing R's property and the chance that he has of retrieving his position by means of this plan becoming an accomplished fact . . . he argues that he will have little interest in labouring for the success of a scheme which will leave him no profit. . . . The people connected with this business [amalgamation] in England have rated Robinson's cooperation so very highly that I should not be doing my duty to them if I did not do everything to secure it in their interests.'

Michell was sympathetic but could do nothing. Robinson could offer no further securities. The shares held by the Cape of Good Hope Bank were improving but not to the extent that would have justified the Standard Bank taking them over. Michell wrote and told Merriman this.

'I do not see how you could take any other course . . .', Merriman replied. 'I much fear that it will end in the C of H Bk sacrificing his shares. . . .' Robinson feared this also. While he was in Cape Town he went to see other money lenders, but without success. He returned to Kimberley and told Merriman that until he could get outside security all his business transactions would have to hang fire. The chances of his getting such security were dim indeed. He was no longer in a position to influence the amalgamation scheme one way or the other.

[4]

The loss of Robinson's support could not have come at a worse time for Merriman. He was on the point of despair by the time Robinson returned from the Cape. His amalgamation scheme looked like going the way of all previous amalgamation schemes.

After a month of negotiations Merriman found himself caught up in the deadly rivalries that strangled so many good intentions in Kimberley. 'The Central people have declined to enter on the discussion,' he told Michell, 'and the English shareholders who pretended to join have evidently been playing a game Kimberley fashion. The De Beers profess that they cannot move without the Central so we have arrived at a sort of impasse.'

This, however, was only part of the problem. To make matters worse, another scheme had been launched which cut right across Merriman's proposals. It had come from an unexpected quarter.

On 13 February 1886, the Kimberley newspapers carried a large advertisement detailing a new plan for amalgamating the mines. In essence it was simplicity itself. The principle mining companies in the four mines were invited to exchange scrip with each other at a commonly acceptable valuation. In this way a joint interest would develop among the various companies which, in effect, would amount to amalgamation. This new plan was sponsored by the De Beers Mining Company and was the brainchild of the company's chairman: Cecil Rhodes.

Later Rhodes was to explain how this particular project came into

being. When he was once in London, he said, he had been shown a
'Memorandum of Association' for a company known as the Globe
Telegraph and Trust Company. This memorandum had set out a
scheme for an exchange of shares which would result in a unified
Trust Company. 'I learned in London', said Rhodes, 'that the
Company I have mentioned has carried out its object most successfully.' He did not say on which visit to London he had been shown
the memorandum, but it seems likely that this was the scheme that
had inspired those folios of intricate figures which he had shown
Philipson-Stow in 1881. If so, he had certainly chosen an opportune
moment to bring his amalgamation scheme forward. It did much to
put paid to Merriman's plan.

Rhodes's intervention came as a surprise to Merriman. Only a
few days earlier Rhodes had supported Merriman and Moulle at
their meeting with the De Beers representatives. As a result of that
meeting Merriman had thought that his plan would be accepted and
that if he could win over the Central Company in the Kimberley
Mine, his work would be finished. This new development put an
end to such wishful thinking.

'Moulle was furious and was for writing and at once breaking off
the whole thing,' Merriman told his wife, 'he had a very warm interview with Rhodes at which some very warm language was used.
Generally we both think that the whole thing is at an end . . . we
have lost confidence ourselves and without that how can we inspire
confidence in others. It is *very* disappointing; lack of success is
nothing but to be fooled as we have been is very annoying. It was
not an enemy either that did us this dishonour. But Rhodes is the
same in business and politics, tricky, unstable and headstrong. Never
able to take a line and follow it! It is a serious defect in his character
and unless he mends it will destroy his usefulness and mar what may
be a fine career . . . I am all the more sorry because I like him
personally.'

This was to be the experience of a great many people in their
dealings with Rhodes. He was a delightful companion and he could,
when he wished, charm anyone into trusting him. But his charm was
deceptive. If he were crossed in business or politics, friendship went
by the board: he would trample on his opponents as easily as he had
captivated them. The only people to whom he was faithful were those
who did not interfere with his plans or those who went along with
them. Those who stood in his way had to go.

Merriman went. Admittedly it was not only Rhodes who had opposed his amalgamation scheme. At a public meeting held in Kimberley on 30 January, Merriman's proposals were unanimously rejected. The various mining companies had also proved refractory. But it was undoubtedly Rhodes who delivered the death blow. Merriman heard later that Rhodes had said quite openly that he had had no intention of allowing the scheme to go through 'and that he did not see why outsiders should interfere in the amalgamation of the mines and so forth'. Merriman had no alternative but to telegraph C. J. Posno, calling the whole thing off. It was quite obvious, as Merriman told his wife, that Rhodes was following a private plan of his own.

[5]

That private plan was not the one advertised in the papers. Rhodes kept his exchange-of-shares idea alive for a few more weeks and then let it drop. It had served its purpose. That he had ever intended it to do more than scare Merriman off is very doubtful. The type of union that would have come about by an exchange of shares was not the type of amalgamation he was seeking. He intended that the mines should be controlled by a single company: that company was De Beers Mining Company, with him at its head.

With Merriman out of the way, Rhodes continued on his old course. By the beginning of 1887 he had taken over all the independent companies in De Beers Mine except one. The last to fall was the Victoria Company, headed by Mr Francis Oats. This company had withstood all Rhodes's overtures and showed no signs of yielding. It was Alfred Beit who came up with a plan for undermining the resistance of Francis Oats. He suggested that his firm should combine with De Beers to buy shares in the Victoria Company in London.

'We felt that if they were bought in the London market', Rhodes explained, 'it would excite no remark. . . . The result of our arrangement was that we did obtain 6,000 shares in the Victoria, jointly, the De Beers Company 3,000 and Beit and Porges 3,000 and they cost us the sum of £57,000, or a little below £20 a share.' They bought the shares at the end of 1886 and kept quiet about them. Negotiations with one or two other companies had to be tied up before they were ready to close in on the Victoria. 'However,' said Rhodes, 'in pursuance of our policy of amalgamation we at last thought the time had arrived to inform the Victoria that we were their largest shareholders . . . and that amalgamation was necessary in our interests.'

That was at the end of April 1887. A few days later, on 6 May, Rhodes was able to report on their success at a general meeting of the De Beers Mining Company. He was thanked by Alfred Beit.

They had good reason to congratulate themselves. By taking over the Victoria Company they had achieved what others had dreamed of: De Beers was the first mine in Griqualand West to come under the control of a single company. Amalgamation had, in part, become a reality.

The next step, of course, had to be in the direction of the Kimberley Mine. This was a much greater challenge. There had been many changes in the Kimberley Mine since the departure of J. X. Merriman a year earlier. Mergers had taken place and a new contestant had entered the Grand Amalgamation stakes. Rhodes now found himself faced by Barney Barnato. Getting rid of Barney was a very different matter from disposing of J. X. Merriman.

CHAPTER FIFTEEN

THE REAL FIGHT BEGINS

THE battle between Barney Barnato and Cecil Rhodes is one of the few episodes in the struggle for control of the diamond industry that has been recorded. But only the last stages of that battle have been dealt with in any detail. Rhodes sketched in the final rounds of the fight in a speech made after it was over; most accounts of it are based entirely on this speech. However, Rhodes did not tell the full story and many wrong assumptions have been made from what he said. For instance, it is usually assumed that he and Barney had been slowly establishing their positions over the years in preparation for the final confrontation. They are pictured as stealthily taking each other's measure before coming to grips. This, of course, fits the popular concept of each side limbering up before entering into a classic battle. Unfortunately, Kimberley contests were rarely of a classic mould; the Rhodes-Barnato clash was not like that at all. It was less of a slow trial of strength and more of a mad scramble for supremacy.

Barney Barnato does not appear to have shown much interest in amalgamation until J. X. Merriman arrived on the scene. It must have been then that he realised J. B. Robinson was no longer in the running. Even after that he made a great show of opposing all plans for unification. This pretended opposition, however, was a matter of tactics. Once he saw how Rhodes had dealt with Merriman, he rose to the challenge. For a while he was forced to keep his hand hidden. He started at a distinct disadvantage.

By 1886 Barney had a large fortune but not much in the way of mining properties. His interests in the Kimberley Mine certainly did not equal Rhodes's holdings at De Beers. In fact, at one time he had considered withdrawing from the mine altogether. Only pressure from his shareholders—particularly Dr Murphy—had persuaded him to keep the Barnato Mining Company alive. His biographers invariably tell of how he stood firm and weathered the depression years. He did nothing of the sort. He was quite ready to throw in his hand.

The Barnato Mining Company, like the Standard Company, was put out of business by the falls of reef in 1882. To keep the company in existence, Barney had to finance it from his own pocket. His pocket was large—he had thriving interests outside the Barnato Company—but he got tired of paying out and getting no return. When, in 1884, he rushed back to Kimberley to help Isaac Joel in the I.D.B. case, he was quite ready to cut his mining losses. 'My principal object was', he claimed, 'to bring the Barnato Company to a definite position one way or the other, as I had been repeatedly called upon to disburse the current expenses of the Company which, considering we had been unable to work up to the time of my arrival for some two and a half years, was virtually bankrupt. Speaking for myself personally, I had really given the matter up in despair of ever seeing our blue ground again.'

As soon as he had recovered from the shock of the Joel case, he set about putting the Barnato Company into liquidation. At a special meeting of the company's shareholders, on 18 August 1884, a resolution was passed dissolving the company. How genuine this move was is difficult to say. In later years Barney was to be accused of dissolving his own companies in order to 'squeeze out' other shareholders. But this could hardly have been the case with the Barnato Company. It was Dr Murphy (recently reinstated with full honours) who proposed the dissolution and it was Dr Murphy and Barney who worked together to save the situation.

Dr Murphy appears to have been the driving force behind their efforts. He was determined not to allow the company to fold without a fight. Barney admitted as much. Addressing a meeting of the Barnato Company some months later, he said: 'I cannot allow this occasion to pass without putting before all those interested in the Company the real person to whom the shareholders owe an ever-lasting debt of gratitude. That gentleman is our Managing Director, Dr Murphy, who, for many months past persistently implored me to abandon and reject the idea of ever parting with our valuable property, no matter what inducements might be offered to me, it being, as is well known, one of the richest portions of the Kimberley Mine. I well remember Dr Murphy's very words when he declared that the future of the Company . . . was contained in the policy of simply displaying a "masterly inactivity" up to the period when we

Corner of the Diamond Market, Kimberley, showing Alfred Beit's first diamond buying office on the left. This is the row of shops upon which Beit founded his fortune

Kimberley Public Library

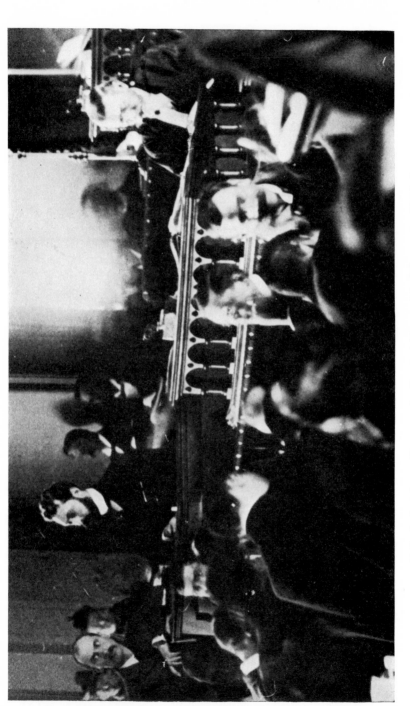

Solly Joel giving evidence at the South African trial of Baron von Veltheim

Barney Barnato
shortly before his death

'Arry. A cartoon of the
flamboyant Harry Barn-
ato in 1906

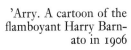

THE PAST

THE PRESENT

OLD CLIENTS

'ARRY.

CLIENTS OF TO DAY

The Snark

Cartoon of Alfred Beit returning from Mashonaland. Lord Randolph Churchill skulks in background

Old before their time, Cecil Rhodes and Alfred Beit
were both in their late forties when this photograph
was taken

Rhodes's coffin arrives at Cape Town Station on its
way to Rhodesia, where he was buried. Draped funeral
train in the background

'A cross between a glorified bungalow and a dwarf Gothic country mansion.' Alfred Beit's London house, 26 Park Lane

Alfred Beit, photographed in the ornate study of his Park Lane mansion

South African Library

Sir Joseph Benjamin Robinson in old age,
wearing his famous pith helmet

could commence operations satisfactorily.' Some of those listening to this paean of praise must have found Barney's effusiveness a little embarrassing. Dr Murphy's unsavoury reputation and past history had, it seems, been hurriedly forgotten.

Strangely enough, one of the causes of their earlier quarrel had helped to strengthen the cynical alliance between Barney and Murphy. Of the many complaints made by Dr Murphy, when he was dismissed as chairman, one was that the Barnato brothers had resolutely opposed the suggestion of an amalgamation with another company in the Kimberley Mine. The company was a neighbouring concern known as the British Company. The Barnatos had wanted nothing to do with the British Company. They had, in fact, packed a specially convened meeting to prevent the suggested amalgamation. 'One of the first acts of the Barnato majority', Dr Murphy had said, 'was to call a special meeting and take from the directors *per se* all amalgamating powers with any other company.' Had amalgamation gone through, it would undoubtedly have benefited the shareholders but it would have broken the grip which the Barnatos exercised over the company. At that time the Barnato Company was a prosperous concern; Barney had no intention of allowing control to slip from his hands. Now, of course, things were different. After dissolution had been decided upon, Dr Murphy revived the suggestion of amalgamating with the British Company. It was a slim hope, but one worth trying.

Barney and Murphy approached representatives of the British Company. There were several meetings; negotiations lasted for five weeks. The Barnato Company was at a disadvantage. Not only were their claims buried, but they had announced their lack of confidence by publicly dissolving the company. Nevertheless, a provisional agreement was reached. At a meeting of the Barnato Company shareholders on 22 October 1884, the resolution dissolving the company was rescinded and amalgamation with the British Company agreed to on condition that 'any deficit which the Barnato Co may have be deducted from the shares given to the Barnato Co instead of remaining a liability on its shares'. They were back in business.

It was probably the resuscitation of the Barnato Company which encouraged Barney to buy six more claims in the Kimberley Mine. These were the claims for which he paid £180,000. He also set to work on his buried claims. Following Robinson's example, he solved his problems by commencing underground mining. The first load of diamondiferous ground to come from the old Barnato claims in three years was hauled on 15 July 1885.

It was a gala occasion. Barney knew how to handle a celebration of this sort. Champagne flowed freely; so did praise. This was the occasion on which he showered Dr Murphy with compliments. Murphy was no less gushing. 'Fortunately for us, gentlemen,' he replied, 'we had in Mr Barnato not only a large shareholder, but a banker to whom the interests of the Company was paramount, for, as I have already said, without his advance and support in keeping the Company afloat, we should long ere have ceased to exist.'

They were both, in fact, charming chaps.

[2]

Shortly after this, rumours of Merriman's amalgamation scheme began to be heard. Barney was in England at the time but his nephews evidently kept him informed. Sensing danger, he must have given instructions for certain precautions to be taken. His first target seems to have been Robinson's Standard Company.

An important factor in Robinson's downfall was the haste with which the Cape of Good Hope Bank started to sell his shares. What caused this panic? One can only surmise. At the time the bank was going through a lean period; it needed only a little pressure to make it jittery. That pressure was undoubtedly applied as far as Robinson's shares were concerned.

Robinson was fully aware of what was happening. At the end of 1885 he visited Cape Town. On his return he took the chair at the annual general meeting of the Standard Company (he did not know it then, but it was to be his last appearance as chairman). He had news for the shareholders. 'I have just returned from Capetown,' he said, 'and there I saw several gentlemen who had received communications from Kimberley and they were themselves thunderstruck with the prospects of the Company as depicted in these letters. The underground works, it was said, were so dangerous that the Government would be compelled to interfere and, in the interest of life and limb, cause them to be stopped. These stories were simply prompted by that green-eyed monster jealousy. These reports that the Company is going to collapse have even reached London, and have emanated from some of our neighbours in the Kimberley Mine . . . that has been the curse of mining in this Province.' Who these neighbours were he did not specify. But he knew only too well that he was being 'squeezed out'. The Cape of Good Hope Bank had begun to sell his shares.

Robinson appears first to have suspected Rhodes and Beit of creating the panic. They were known to be actively working for the control of the mines and this made them natural suspects. 'One day he came to my office,' said Beit, 'and said that he had lost all his money; that Rhodes and I had ruined him.' This is what Robinson thought in the middle of 1886. Had he waited a little longer he might have changed his mind. At the next annual meeting of the Standard Company, it was neither Rhodes nor Beit who emerged as the largest shareholder: it was Barney Barnato.

For the first time since the formation of the company Robinson was not in the chair; he was not even at the meeting. The chair was taken by a Mr Pistorius, but this was merely a formality. The longest and most authoritative speech was made by Barney Barnato. He lectured the shareholders on the running of the company and urged them to agree to the recent proposals for an amalgamation with the Barnato Company.

This was his answer to Rhodes's moves in the De Beers Mine. The combination of his own claims and those of the Standard Company would put him in a powerful position in the Kimberley Mine. And, like many others, he was convinced that whoever controlled the Kimberley Mine would eventually control the diamond industry.

Barney's *coup* was not accomplished without a skirmish. Some of the Standard shareholders put up a fight. The reputation of the Barnato brothers did not inspire confidence; they viewed Barney's sudden arrival in their midst with loudly voiced suspicion. For as long as they could, they held out against the amalgamation; but Barney proved too strong for them. The annual general meeting was held on 13 December 1886; two months later another meeting was called. This second meeting was a much larger affair: Barney had seen to that. So many new shareholders turned up that it was necessary to hire the local auctioneer's rooms to accommodate them. Anyone who had attended meetings of the Barnato Company would have recognised many familiar faces: Harry and Barney Barnato were there, so was Woolf Joel, so was Dr Murphy, and so were all Barney Barnato's friends and acquaintances in Kimberley. Thirty of the shareholders attending also held proxies for 5,000 votes. The proceedings were brisk. Woolf Joel was elected to the board of the

Standard Company and the proposal for a merger of the Standard and Barnato Companies was carried unanimously to loud applause. The new concern took the name of the larger of the two companies. For the short period of its existence, it continued to be known as the Standard Company.

[3]

Barney next turned his attention to the most important company in the Kimberley Mine: Baring-Gould's Central Company. This was the company which Merriman had regarded as vital to the success of his unification scheme. His failure to win over the Central shareholders had placed him at the mercy of Rhodes. Barney took no chances. Some of the more influential shareholders in this company were former diggers who had left the diamond fields and were now living in great style in London. Immediately the amalgamation with the Standard was tied up, Harry and Barney left for London. Woolf Joel remained behind to look after things in Kimberley.

It did not take long to bring the new powerful Standard and the Central together. Events were speeded up by Barney's recent *coup*; there were few people in Kimberley who did not recognise the way things were going. The Barnatos could no longer be ignored. At the end of June 1887, Woolf Joel sent Barney a cable to say that negotiations in Kimberley were coming to a head. Barney rushed back to the Cape in pretended alarm.

'When I left London,' he said, 'having received a cable that amalgamation was being contemplated and certain arrangements having been submitted to me, I must say I was heartily opposed to it and wired to that effect. My nephew immediately cabled me asking me to come out at once. There were other shareholders in both concerns—Central and Standard—in London at the time, and when they heard that I was opposed to the scheme of amalgamation they came to see me with the object of impressing upon me the advantages to both concerns even without knowing the details of amalgamation. They pointed out to me the advantages which the De Beers Mine had derived by the amalgamation of properties. I abstained from giving an opinion until I arrived here. When I did arrive and saw the terms of agreement, and had spoken to the directors, everyone seeming to think the project was a good and sound one for both Companies, I did not care about throwing obstacles in the way.'

Barney made it seem as if the idea of amalgamation had been

presented to him out of the blue. The events of the last eighteen months might never have happened. But the role of reluctant amalgamator did not suit him. He played it as drama when it was obviously farce. The terms of the proposed agreement might not have pleased him (they rarely did) but he was fully aware of the urgency of this all-important merger. He did not need the shareholders of the Standard and Central to direct his attention to what was happening in the De Beers Mine. His show of hesitancy was simply a blind. He wanted to disguise, if he could, the eagerness with which he was consolidating his interests in the Kimberley Mine. There could not have been many who were fooled.

The amalgamation of the Standard and Central Companies was formally decided upon at two meetings held on 7 July 1886. In the morning, the Standard shareholders met to consider a proposal that the company be put into liquidation in order to amalgamate with the Central. There was very little discussion and the proposal was carried unanimously. The meeting of the Central shareholders that afternoon was a good deal livelier but the result was the same.

By the end of the day the Standard Company no longer existed and the new Central Company was in possession of the greater part of the Kimberley Mine.

Barney attended both meetings. The result undoubtedly delighted him. In little over a year he had accomplished what it had taken Cecil Rhodes the best part of six years to achieve. At the beginning of 1886 there had been four powerful companies in the Kimberley Mine: the Central, the French, the Standard and the Barnato. Of these four the Barnato had been the smallest. Now, only the French Company and a few minor concerns stood between Barney and complete control of the richest diamond mine in Griqualand West.

For there could be no doubt that Barney Barnato was the sole power behind the latest amalgamation. Francis Baring-Gould was still nominally the chairman of the Central Company but the Barnato family held the majority of shares; it was Barney who pulled the strings. Anyone attempting to amalgamate the four mines would have Mr Barnato to deal with. Before long all Kimberley was made aware of this.

At the conclusion of the Central Company's meeting, a series of loud explosions were heard in the vicinity of the Kimberley Mine. People passing by were said to be startled out of their wits by the noise. It sounded, claimed the *Diamond Fields Advertiser*, as if 'the

Transvaal Navy had suddenly appeared and were bombarding dear old Kimberley. Hundreds of people rushed towards the mine, where the cannon, that is the dynamite shots, were being fired all round the edge, while the Companies flags were seen waving from the Central and Standard Companies works. It was an "Amalgamation Salute" and it sounded quite gay and joyful, but nervous persons did not like it at first.'

One nervous person would not have liked it at all. That was Cecil John Rhodes. Those dynamite shots were aimed right at the heart of his great vision. Fortunately he did not hear them. He had sailed from Cape Town the day before. He was on his way to England to organise a counterblast.

CHAPTER SIXTEEN

THE BATTLE OF THE GIANTS

IN the early days, when the Kimberley Mine was still known as New Rush, there had been considerable doubt as to the mine's potential. The first finds there had been staggering but it was thought that they were too good to last. In those days, of course, diamonds in all four mines were expected to peter out after a depth of thirty to fifty feet. In the opinion of many diggers, this put a limit to the length of time that New Rush could be worked. The stampede to Colesberg Koppie had resulted in such overcrowding that it was confidently expected that digging there would come to an end sooner than in the other mines. 'Such an immense number of hands are now at work at the Colesberg Kopje,' wrote Charles Payton in 1871, 'that many people say it will be entirely worked out in from six to twelve months' time.' Nor did the diamonds found at New Rush provide much encouragement. Plentiful they might be, but they were of an inferior quality. Many were of a bad shape and size and the larger stones, over 10 carats, tended to be off-colour. New Rush, in fact, was regarded—in diggers' terms—as a flash in the pan.

Hope for the diamond industry had centred on Dutoitspan and its neighbour Bulfontein. Finds in these mines were not so spectacular but they were more steady and rewarding. Dutoitspan had first attracted diggers to the dry diggings and it was from this mine that a prosperous and permanent industry was expected to develop. 'The centre of a country is that point to which the most and largest interests converge,' it was said. 'This point in Griqualand West, is Dutoitspan, and, so far as we can see, will continue to be. . . . Every business man of the colony feels sure that Dutoitspan is destined to be an important centre.'

But it had not worked out like that. Once the arid 'yellow' diggings had been exhausted and mining of 'blue' ground had begun, it was realised that Colesberg Koppie was even richer than had been imagined. New Rush had become the Kimberley Mine and it was between this mine and the neighbouring De Beers Mine that the town of Kimberley had developed. Dutoitspan and Bulfontein continued to yield quantities of diamonds but work in these mines proceeded at a slower pace; they were considered of secondary importance to Kimberley and De Beers.

In the race towards amalgamation, little heed was paid to either Dutoitspan or Bulfontein. Most people spoke of them as the 'poorer mines'. They would, of course, play a part in the final stages of the game but they were not essential to the success or failure of amalgamation. Rhodes was later to sum up the general attitude towards these two mines. 'I read a very good story the other day', he said, 'describing a class of mine which is very common in the mining world. "This mine", the writer says, 'is just one of those mines that are too rich to leave and too poor to pay." I thought a good deal over that in connection with our poorer mines. If I said anything about them, more especially Dutoitspan, and in a lesser degree Bulfontein, they would be described by me as too poor to pay and too rich to leave. They have to be considered as a producing factor, but as a paying factor they are nowhere.' Time and events would decide their fate.

Therefore everything hinged on the Kimberley and De Beers mines. More specifically everything hinged upon the Kimberley Mine. Rhodes had more or less tied up De Beers; Barney Barnato was still a few pieces missing in Kimberley. The most important of these pieces was the highly desirable *Compagnie française des diamonds du Cap* or, as it was colloquially known, the French Company. This company, founded in 1875 by Jules Porges and represented in Kimberley by his partners, Wernher and Beit, now became the focus of attention. It occupied an important strategic position, both financially and geographically. A line of its claims ran across the Kimberley Mine, from north to south, dividing the holdings of the Central from those of the former Standard. So long as these claims remained outside the Barnato net, the amalgamation of the Central and Standard was, to all intents and purposes, a financial union only. Barney knew this. Rhodes knew this. All Kimberley watched and waited for the next move.

[2]

Rhodes had a slight edge on Barney. The French Company was represented by Wernher and Beit and Beit was Rhodes's ally. It was not much of an edge. Beit only represented the French Company, he did not control it; but at least he was able to keep Rhodes informed of any moves made against him.

It would have been more to the point if Rhodes had had a definite foothold in the Kimberley Mine. This is what he sadly lacked. He had been so busy manipulating his holdings at De Beers that he had allowed Barney to entrench at Kimberley without any real opposition. If Rhodes had held a few of the smaller properties in the Kimberley Mine, he might have been able to hold Barney at bay for a while. As it was the field was wide open. Barney had only to snatch the French Company, as he had snatched other companies, and he was home and dry: none of the smaller concerns was strong enough to stand out against him.

Rhodes had realised this. A few weeks before Barney finalised the amalgamation of the Central and Standard Companies, Rhodes had made a last minute bid for a block of claims in the Kimberley Mine. These claims belonged to W. A. Hall. They stretched into the centre of the mine and, next to the French Company's holdings, they were probably the most important claims outside Barnato's reach. It was sometime in May 1887 that Rhodes tried to buy the W. A. Hall Company, only to find it was too late. He was outbid by a syndicate headed by Sir Donald Currie, the shipping magnate, who also had thoughts of amalgamation.

Immediately after securing the W. A. Hall shares for £110,000 Sir Donald left for England. Two of Rhodes's agents joined him on his voyage home. They had been instructed to make an all-out effort to persuade him to re-sell the shares. It seems they did a good job. Sir Donald listened to them and was half tempted by Rhodes's offer. However, when the ship reached Lisbon he discovered that the market value of the shares was higher than the price that Rhodes was proposing. This put an end to all negotiations. 'You young thieves,' he said to the agents, 'had I listened to you I would have sold at a loss.' Somewhat disconcerted, the young men cabled to Rhodes to tell him of Currie's decision. Rhodes sprang into action. By the time the ship reached Plymouth, a day or two later, he had beared the market to such effect that the shares were no longer worth what his agents had offered for them. Sir Donald had been taught a lesson. Although he is said to have held on to his shares for a while longer, he dropped out of the competition.

This little exercise in financial politics did not get Rhodes much further. It was now June and the amalgamation of the Central and Standard was almost complete. He had to move quickly. His only hope of blocking a complete take-over by Barnato was an outright purchase of the French Company. This presented a formidable problem. As the battle for control of the mine had intensified, so had the French Company's shares risen. Rhodes was a rich man but his money was tied up in his holdings; he dare not release his grip on any of the shares he held. He needed extra capital. He needed to negotiate a very large loan.

Alfred Beit came to his rescue. By using his influence with French and German financiers, he obtained the promise of a substantial loan in exchange for a block of De Beers shares. By farming out the shares in Europe, there would be little danger of them falling into Barnato's hands. It was a beginning, but it was not enough. Then Beit suggested that they approach Rothschilds of London. As the largest financial house in Europe, Rothschilds had both the name and the resources to give the backing they needed. Rhodes leaped at the suggestion. Rothschilds had already shown an active interest in the South African diamond mines and, in 1875, they had advanced Disraeli no less than four million pounds for the purchase of shares in the Suez Canal. With the House of Rothschild behind him, Rhodes would be more than a match for any move made by Barney Barnato.

Beit was able to provide an introduction to Lord Rothschild but Rhodes left nothing to chance. He brought all the influence he could to bear. At the beginning of May 1887, he had engaged a new manager for the De Beers Mining Company. This was Gardner Williams, an experienced American mining engineer. Williams was well known in mining circles; among his friends was E. C. De Crano, the engineer who advised the Rothschilds on South African mining affairs. Rhodes asked Williams to send De Crano a full report on the diamond industry, stressing the importance of amalgamation, and try to persuade his friend to put the scheme for purchasing the French Company to Rothschilds. Williams did this and Rhodes then left Kimberley for Cape Town to attend the parliamentary session.

Hardly had Rhodes arrived in Cape Town before Barney Barnato returned to South Africa to finalise the amalgamation of the Central and Standard Companies. This appears to have thrown Rhodes into a panic. He immediately wired Gardner Williams, asking the engineer to join him in Cape Town, and on 6 July they sailed for England.

The following day Barney startled Kimberley with his Amalgamation Salute.

On reaching England Rhodes lost no time in contacting Lord Rothschild. A meeting was arranged and the plan for buying the French Company discussed. The Rothschilds had already studied the report which Gardner Williams had sent to De Crano and had been favourably impressed. All Rhodes had to do was to fill in the background. There is no record of what Rhodes told Lord Rothschild, but Barney Barnato was later to give his own version of the interview. According to Lou Cohen, Barney liked to claim that Rhodes told Lord Rothschild that the only man he feared in South Africa was 'a cunning little Jew called Barnato'. The idea of frightening Rhodes evidently appealed to Barney. He delighted in telling this story. 'I can see Barney now as he told me . . .', says Cohen. 'His cold blue eyes beamed with stony merriment, and he laughed loudly and heartily as he described what he considered the honour done him.'

Whether Lord Rothschild was similarly amused, one does not know. What is more certain is that Rhodes managed to win Rothschild over. 'Well, Mr Rhodes,' Rothschild said at the end of the meeting, 'you go to Paris and see what you can do in reference to the purchase of the French Company's property, and in the meantime I will see if I can raise the £1,000,000 which you desire.' The promise was vague but encouraging.

As Rhodes and his party were leaving the room, however, Lord Rothschild stopped De Crano. 'You may tell Mr Rhodes,' he said, 'that if he can buy the French Company, I think I can raise the million pounds sterling.' This, in the language of high finance, put things on a much firmer footing. It gave Rhodes all the reassurance he needed. That same evening he left for Paris to negotiate with the directors of the French Company.

Rhodes loved to boast of the swiftness with which he pulled off this deal. 'You know the story of my getting on board the steamer at Cape Town, going home and buying the French Company within twenty-four hours,' he used to say. As so often happened, he was taking more credit than was his due. There can be little doubt that he owed much (if not all) of his success in dealing with the directors of the French Company to Alfred Beit. For Beit had already prepared his associates in Paris for Rhodes's coming. He had persuaded Jules Porges that amalgamation of the mines was a sound financial move and that Rhodes was the man to back in bringing it about. Porges, in turn, had recognised the need to stabilise the diamond market and, as the founder of the French Company, had used his influence with his fellow directors to get Rhodes's provisional offer accepted. It was, in fact, thanks to Porges and Beit that the transaction went through with such speed; but they were content to remain in the background and draw their dividends while Rhodes paraded his genius before an admiring public.

The French directors agreed to the sale of the company for £1,400,000. To meet their price, Rhodes took up a loan of £75,000 from Rothschilds and issued 50,000 De Beers shares—at £15 per share—to the French and German financiers whom Alfred Beit had sounded out earlier. (These shares were issued on the understanding that De Beers and the financiers would divide between them any profit from the rise in value of the shares. Soon they rose to £22. Each side made a handsome £100,000 from the deal.) The sale of the company was only provisional, however. It had to be confirmed at a meeting of the French shareholders, which was called for October. This, in a way, was merely a formality. The Trust Deed of the company gave the directors power to sell the property, provided their decision was a majority one. As the majority of the directors were in favour, any opposition at the shareholders' meeting could easily be overcome.

In banking on the simplicity of this procedure, Rhodes had reckoned without Barney Barnato.

[3]

Barnato, of course, had kept himself informed on what was happening. He was not prepared to let the French Company slip through his fingers without a fight. Already he had prepared himself for battle. By employing his usual tactics, he had acquired a susbstantial interest in the French Company; he held no less than a fifth of the company's shares. When he heard of the deal Rhodes had negotiated, he lost no time in countering it. He put in a bid of his own for the company. He topped Rhodes's offer by £300,000. The Central Company, he said, was prepared to pay £1,700,000 in cash for the French Company. This, together with the shares he held, put him in a strong position.

But it was not quite strong enough. Rhodes's allies in Paris stood by him. Barnato was told that a provisional agreement had been negotiated and that he, like the other shareholders, would have to abide by it. Needless to say, Barney was not prepared to do any such thing.

On 21 September, he called a meeting of all the Kimberley shareholders in the French Company. It was a mammoth gathering. Once again it was necessary to hire the local auction rooms. Barney took the chair. He was in a fighting mood.

'They had all more or less come out to this country,' he was reported as saying, 'not for the benefit of their health, but for the benefit of their pockets, and he would ask them to consider seriously the matter which was before them.' He went on to denounce the proposed sale of the company. With a wealth of facts and figures, he explained how the take-over would prove detrimental to the smaller shareholders. The company was being sold, he said, by people who had no real knowledge of the worth of the Kimberley Mine. 'They must take into consideration', he declared, 'that the purchase of the French Company's property, involving more than a million of money, was settled in Paris in less than twenty-four hours after the purchaser arrived there, and he asked them how on earth they could possibly have taken into consideration the value of the property?' This was a matter which, in his opinion, should be open to debate among the shareholders of the company. It was up to them to make their voices heard at the meeting which was to take place in Paris the following month. They had a good case for opposing the majority decision of the directors. He had consulted his lawyers and discovered that if it could be proved that those directors stood to gain by the sale, even if only by the extent of one share, the shareholders could hold them responsible for 'every sixpence of loss' resulting from the deal.

He was enthusiastically supported by the ubiquitous Dr Murphy. The good doctor had also come to the meeting armed with facts and figures. He proceeded to detail the various ways in which the French directors had neglected the interests of the Kimberley shareholders. 'Were they', he asked, 'to be allowed, as they had done in the past, to deal with the shareholders' property in any way which they thought proper, and as a climax to all this sell the property without consulting the shareholders, after twenty-four hours deliberation?'

Both Barney and Murphy were heartily applauded. The only real attempt at opposition came from Alfred Beit. Unfortunately, the painfully shy Beit was so nervous when he got up to speak that very few heard what he had to say; those who did hear must have been puzzled by his highly involved financial arguments. His objections carried little or no weight.

At the end of the meeting, Dr Murphy proposed that a fund be set up so that the shareholders could secure legal assistance, both in Kimberley and France. A cable was sent to the directors in Paris protesting against the sale to De Beers 'as a better offer has already been made by the Kimberley Central Company'. It was clear that Cecil Rhodes was in for some pretty stiff opposition.

Rhodes was quick to appreciate the danger. He had just arrived back in South Africa and was confidently awaiting the approval of the French shareholders. Now he was forced to change his tactics. Rothschilds, alarmed at the prospect of being involved in a Kimberley financial brawl, requested him to approach Barnato personally. He did as he was asked. Barney was all smiles, but he refused to compromise. 'It might be good enough for me,' he explained, 'but what about all the other shareholders who look to me to get them a good price for their shares?' Rhodes saw this as a show of strength, rather than an expression of concern. He tried to appeal to reason. 'You can go and offer £300,000 more than we do for the French,' he said, 'but we will offer another £300,000 on that; you can go on and bid for the benefit of the French shareholders *ad infinitum*, because we shall have it in the end.'

Barney remained unimpressed. In any game of beggar-my-neighbour he felt he was as strong as Rhodes. For one thing he was much richer. Rhodes had borrowed heavily to make his initial offer, he could not compete with the Barnato group indefinitely. If anyone had to give in it would not be the Barnatos.

Rhodes knew this as well. He was wise enough not to attempt to better the Barnato offer. Instead, he graciously bowed out of the race. At least, that is what he appeared to do. To everyone's surprise (Barney's not the least) he agreed to hand over the French Company. He made Barney a splendid offer. He said that if he was allowed to purchase the French Company without interference, he would re-sell it to the Central for £300,000 cash and the remainder in Central shares. Barney was delighted. The deal was closed.

At a meeting of the Central Company's shareholders, on 3 November 1887, Barney was able to boast of his success. With great glee he explained that he had originally offered 70,000 Central shares for

stand Barney. In later years, nothing annoyed him more than to be linked with Barnato. 'Mr Beit did not seem to wish to talk of Barney Barnato,' said an interviewer; 'he admitted his gifts, but evidently did not like him, and contented himself with saying that he never knew him well till the great De Beers amalgamation scheme brought them together in 1887.' Alfred Beit was too honest to disguise his feelings.

Behind the friendly façade, each tried to destroy the other. Rhodes started it; Barnato followed suit. They bought up shares with reckless determination. And they bought at a time when diamonds were at an all-time low; the prices they fetched hardly covered the cost of production. The value of Kimberley Central shares soared. When the struggle began they stood at £14, soon they had risen to £49. This created new problems.

Some of Barney's fair-weather supporters could not resist the temptation to make a quick profit. As the value of the shares rose, they stopped buying for Barnato and started selling for themselves. Rhodes was quick to take advantage of this disloyalty. He warned Barney what was happening. They met in the street one morning and Rhodes said: 'I'll tell what you will find out presently, and that is you'll be left alone in the Central. . . . Here you have your leading shareholders patting you on the back and backing you up, but selling out around the corner all the time.' Barney would never admit it, but this, as much as anything, proved his undoing.

For the most part Rhodes's allies stood firm. Only once was he unintentionally let down. The defection, it is said, was caused by Alfred Beit's enthusiasm. Beit had been buying all the Central shares he could lay hands on without telling Julius Wernher. Finally, he went to confess himself. 'Oh, that's all right,' said Wernher. 'I found the firm was getting more Kimberley shares than I liked, so I have sold a lot at excellent prices.' This was the one exception. In the long run, Beit proved to be Rhodes's staunchest backer. He and his associates are said to have secured some 23,000 shares between them. They held on to all but a handful.

At a crucial stage in the battle, it was rumoured that one of Rhodes's European backers—probably Rothschilds—were thinking of withdrawing support. The crisis showed Rhodes at his best. Meeting Beit in the Kimberley Club for a drink, he asked a steward to bring him half a dozen promissory note forms. He signed and folded them and handed them to Beit. 'You've staked everything on the success of this in backing me,' he said. 'I do not know how it will end now that these people have gone back on us, but I want you to take these. They are signed in blank. Whatever I have is yours to back you if you need it.' The rumour proved false and the crisis resolved itself. Four years later Beit found the signed promissory notes at the bottom of a despatch box and burned them. Such was his trust in Beit that Rhodes had forgotten that he had signed them.

Conflict with Barnato could not last indefinitely. The low price of diamonds and the high price of shares proved too much for Barney. He was forced to capitulate. Rhodes had approached him with offers throughout the struggle; finally he yielded. By this time Rhodes had increased his holding in the Kimberley Central from one fifth to three fifths. There was little point in Barney holding out much longer. In March 1888 he accepted Rhodes's terms. He agreed to give up the Central in exchange for De Beers shares.

Various stories are told about Barnato's surrender. Most of them are doubtful. For instance, it is said that Barney was swayed by Rhodes's promise to ensure his election as a Member of the Cape Assembly and to make him a member of the Kimberley Club. Barney *was* elected to the Cape Assembly with Rhodes's help, but this could have been only a minor consideration. There can be little doubt that Barney was quite capable of ensuring his own election. He had the money and he had influential friends in the Government. His name had been suggested as a candidate in several elections and few doubted that he could win if he chose to stand.

The promise of membership to the Kimberley Club could have been of even less consequence. According to popular belief, Barney had been barred from joining the club because an anti-semitic rule prevented him becoming a member. But the club had Jewish members; if Barney was excluded it was not because he was Jewish but because he was Barney Barnato. It is unlikely that this worried him unduly. Undoubtedly he had social ambitions, but he was too hard-headed to allow such considerations to influence a business deal. Rhodes was not the only member of the Kimberley Club and, had Barney been determined to join, he would undoubtedly have found ways and means of doing so. To think otherwise is to underestimate Barney Barnato.

However, the most colourful story concerns both the Kimberley Club and another popular legend. It is said that, after introducing Barney to the Kimberley Club, Rhodes made a request. 'Well, you've had your whim,' Rhodes is supposed to have said; 'I should like to have mine, which you alone in Kimberley can satisfy. I have always wanted to see a bucketful of diamonds; will you produce one?' This is thought to have flattered Barnato. It is claimed that he shovelled all his available diamonds into a bucket and that Rhodes, plunging his hands into the bucket, 'lifted out handfuls of the glittering gems and luxuriously let them stream back through his fingers like water'.

It makes a colourful picture, but it is most unlikely. Had Rhodes wanted to indulge such a whim, he had had plenty of opportunities for doing so. There had, for instance, been the occasion when he had poured his own diamonds into a bucket to thwart Barney. Perhaps this is how the story originated. Over the years Rhodes's 'historic trick' has become confused with Rhodes's historic *coup*. Buckets full of diamonds are apt to dazzle the imagination.

One would like to believe that the deal was brought to such a glittering climax. It would have made a fitting end to one of the greatest financial transactions in history.

CHAPTER SEVENTEEN

THE VICTORS

IN the middle of his financial duel with Barney Barnato, Rhodes was obliged to attend to some urgent political business. It concerned his plans for central Africa.

With Bechuanaland firmly protected by Britain, he had more or less secured his 'Suez Canal' to the interior. Once this had been achieved he had begun to contemplate taking further steps northwards. He had turned his attention to Matabeleland, the African-ruled territory north of Bechuanaland. Unfortunately, his were not the only eyes turned in that direction.

Rumours of the mineral potential of Matabeleland had already attracted gold-prospectors to the territory. The Matabele chief, Lobengula, had been pestered for some years by concession-hunters asking to mine his lands. These fortune seekers had not made much headway. Lobengula was wary of allowing the white man into his domains. His concern was not so much for the mineral wealth of Matabeleland as for his rights as the ruler of the territory. The history of southern Africa was such as to make any black ruler fear the approach of white men, whatever their excuse for coming. Consequently, Lobengula had severely restricted the activities of those who came looking for gold.

Rhodes knew this. He knew that the time would come when he would have to deal with Lobengula and with the concession-hunters. But he had not been unduly worried. He intended tackling that particular problem when he came to it. First it was necessary for him to complete the amalgamation of the diamond mines. Not until he had sufficient money behind him could he risk further experiments in territorial advancement.

Then, at a time when he needed to give full attention to his struggle with Barney Barnato, Rhodes's political plans were thrown out of gear. In December 1887— shortly after the amalgamation of the French and Central Companies—Rhodes heard that an emissary of the Transvaal Government had negotiated a treaty of friendship with Lobengula. In itself, the treaty amounted to very little: friendly messages were exchanged, the Transvaal was granted the right to establish a consul at Lobengula's kraal and given powers of jurisdiction over Transvaal subjects living in Matabeleland. However, Rhodes was told that the Boers in the Transvaal intended to use this treaty to check British enterprise north of Bechuanaland. Their consul was about to start for Lobengula's kraal; already they were claiming the right to take up concessions in the country. Vague as the news appears to have been, it was sufficient to alarm Rhodes. He felt compelled to take action.

Whether, as is sometimes asserted, he arranged for the Transvaal consul to be murdered is not certain. Piet Grobler, the consul, was killed on his way back from Matabeleland some months later and Rhodes was suspected of being behind the killing. President Kruger of the Transvaal is said to be one of those who were convinced that Rhodes's agents were responsible for Grobler's death. However, the evidence for this is slight. The Boers (not without reason) were willing to believe anything bad about Cecil Rhodes.

There is, however, no doubt about Rhodes's own action on receiving the news. Unable to visit Matabeleland himself, he did the next best thing: he appealed to the 'Imperial factor'. He made a quick dash to Grahamstown, in the Eastern Cape, where the British High Commissioner, Sir Hercules Robinson, was spending the Christmas holidays. Bursting in upon the startled Sir Hercules, Rhodes demanded that a protectorate be declared over Matabeleland and its subject state, Mashonaland. Bechuanaland had been safe-guarded in this way and there seemed no reason why the move should not be repeated. But Sir Hercules was hesitant. It was one thing for the enthusiastic Rhodes to expect nations to fall into his lap at the stroke of a pen but quite another for Her Majesty's representative in South Africa to explain away such actions to the British Government.

But it was impossible to refuse Rhodes entirely. At times like this, when he set out to get his own way—to win over someone 'on the personal'—Rhodes's powers of persuasion were prodigious. He might find the going difficult with a shrewd financier like Barney Barnato, but he invariably won over those who were less self-interested. The poetic grandeur of his vision, his irrepressible enthusiasm and engaging candour, the very single-mindedness of his arguments could mesmerise the most hard-headed of politicians; few were known to resist him for long. Sir Hercules was no exception. Although he would not annex Lobengula's territory outright, he was prepared to taken an option on it. At Rhodes's request he sent a despatch to Matabeleland asking Lobengula not to enter into any treaty with a foreign power or to part with any land without the sanction of the British High Commissioner. To this Lobengula agreed.

On 11 February 1888, a treaty of 'peace and unity' between Victoria, the Great White Queen of England, and Lobengula, the Mighty Elephant of Matabeleland, was signed at the chief's kraal. This was not quite what Rhodes had hoped for but it was enough. It gave him the breathing space he needed.

On his return to Kimberley, Rhodes set about putting his relationship with Lobengula on a firmer footing. The Transvaal had been warned off, but there were still the concession-hunters to contend with. These private speculators had to be cleared from the field; the only way of getting rid of them was for Rhodes to obtain an overriding concession to mine minerals in Matabeleland. This became his new objective. Beit, already a convert to Rhodes's brand of romantic imperialism, threw his financial weight behind the new venture. Before the treaty between Lobengula and Britain was signed, Rhodes and Beit had despatched a representative to Matabeleland with the object of obtaining a gold-mining concession. Unfortunately, the man they chose for the job fell ill and was forced to return without accomplishing anything. This did not deter Rhodes. He set about organising a stronger and healthier mission to the Matabele chief, under the leadership of his old partner Charles Rudd. This deputation had more success. After months of negotiations, Rudd obtained the all-embracing mineral concession. Rhodes was thus established in Matabeleland and Mashonaland—the two countries which were to bear his name when they became Rhodesia.

[2]

The unexpected flare-up of the Matabeleland issue, had a profound effect on Rhodes's final settlement with Barney Barnato. He had always intended to use the amalgamated diamond mines to finance his imperial vision; now, the need to secure that financial backing had become imperative. Rhodes insisted that the Trust Deed of the new company should provide for the accomplishment of his political ambitions. Barney objected. He was a businessman and his business was diamonds, not politics. They argued the point for several days. In the end, Rhodes had his way.

The final meeting to discuss the Trust Deed was held in Rhodes's corrugated iron cottage. Barney brought Woolf Joel with him, Rhodes was backed by Alfred Beit. The four of them argued throughout the night. They became tired and irritable, but neither side was prepared to give in. Rhodes won eventually because he had to. For him this final stage of the battle was the most vital. Amalgamation meant little or nothing if he could not use it for his own ends. He brought all his powers of persuasion to bear on Barney. He produced facts, figures and maps to support his arguments. He appealed to Barnato's business instincts as well as to his imagination. There was no telling what riches might be found in central Africa, he contended. Barney owed it to himself as well as to the new company to make provision for exploiting those riches.

'Aren't those just dreams of the future?' asked Woolf Joel. 'Dreams don't pay dividends.'

'No my friend,' replied Rhodes, 'they're not dreams, they're plans. There's a difference.'

Night gave way to morning. They were all exhausted. At last Barney, struggling to keep awake, gave in. It was then that he made his famous submission. 'Some people', he shrugged, 'have a fancy for *this* thing and some for *that*; *you* have a fancy for making an Empire. Well, I suppose I must give it to you.'

The new amalgamated concern was called the De Beers Consolidated Mines. It was incorporated in March 1888 with a modest capital of £100,000 in £5 shares. The Trust Deed, which was drawn up later, gave Rhodes all he needed. Certain demands made by Barney had also been met. One of his objections to amalgamation had been that shareholders of the new company might resent his presence as a director and refuse to re-elect him. He had good reason to think this. He was far from popular with the men who supported Rhodes. F. Philipson-Stow, for instance, had declared: 'The day that man comes onto the Board, I go off; and there will be others with me.' It had taken the combined efforts of Rhodes and Beit to get Stow to change his mind. (When Barney heard that it had been necessary to talk Stow round, he registered no surprise. 'Yes,' he observed acidly, 'I thought I must have been holding things up.') It was to

counter such objections that Rhodes accepted Barnato's suggestion
of constituting life-governorships for the new corporation. Four
life-governors were appointed—Rhodes, Beit, Barnato and Philipson-
Stow—and between them they held all but twenty-five of the De
Beers Consolidated Mines original shares. They reserved to themselves the power of issuing further shares.

The scheme of life-governorships was not popular. In April
1888, Philipson-Stow was despatched to London to explain the
purpose of the Trust Deed to Lord Rothschild and the British shareholders of the old Central Company. He was empowered to offer
Baring-Gould a fifth life-governorship. His offer was refused. 'There
was considerable bitterness', he said, 'about Mr Barnato having
made terms for himself personally with Messrs Rhodes and Beit
wholly regardless of the promises he had given to Mr Baring-Gould
and the Kimberley Directors to work with them in the interests of
the general body of shareholders and do his best for all.'

Much of this bitterness had been occasioned by Barney's high-
handed assumption of life-governorship. The British shareholders
felt that they had been betrayed and the London financial press
denounced the scheme. Baring-Gould refused to compromise
himself by following Barney's example. 'He said', wrote Stow,
'offers had been made to him personally direct from Kimberley on
behalf of the De Beers people to give him the same terms as Mr
Barnato but that he, Mr Baring-Gould, had declined them unless the
outstanding Central shareholders were included.'

Such integrity was commendable, but of little avail. Nothing could
withstand the powerful combination of Rhodes, Beit and Barnato.
They had fought the amalgamation battle and they dictated the
peace terms. On 31 March 1888, Rhodes had outlined each stage of
his struggle with the Central Company to the shareholders of De
Beers. The speech was one of the longest—it ran to nine thousand
words—and one of the most remarkable he ever made. He spoke
without notes and, for the most part, held his audience enraptured.

Barney Barnato was one of the few who did not succumb to
Rhodes's spell on this occasion. While Rhodes spoke Barney sat
smiling. Rhodes was unable to ignore him. 'These are the facts, I
can assure you,' he insisted, 'although Mr Barnato may shake his
head and smile.' But, as Barney well knew, he was not presenting all
the facts. There were some things better left unsaid. And Barney,
although obviously cynical, did not interrupt or contradict him.

They were allies now. Uneasy allies perhaps, but allies nevertheless. Barney had a new respect for Rhodes. That all-night session, haggling over the Trust Deed, had taught him a lesson. In later years he carefully avoided Rhodes when they differed. He knew the risk he ran by arguing. 'When you have been with him half an hour,' he would say, 'you not only agree with him but come to believe you have always held his opinion. . . . You can't resist him: you must be with him.'

Rhodes was delighted with the success of his speech. When he had finished, the shareholders crowded round to congratulate him. He was too elated to listen. Pushing his way through the crowd, he beckoned to Alfred Beit to follow him. He strode towards the De Beers Mine, with little Beit scampering behind him. At the mine, they were quickly surrounded by African mine workers. Rhodes asked Beit if he had any money with him and, when Beit handed him a bag of fifty gold sovereigns, scattered the coins magnanimously among the laughing labourers. Still heady with success, he turned and walked silently towards the Kimberley Club. On the way he suddenly stopped and waited for the panting Beit to catch him up. 'And tonight', he declared cynically, 'they will talk it over with their wives and tomorrow they will sell like hell!' Beit never got his fifty sovereigns back.

Towards the end of his speech, Rhodes had referred to the 'poorer mines': Dutoitspan and Bulfontein. He said that De Beers intended to make them fair offers, which they would be wise to accept. 'If they do not,' he warned, 'there may be a period of antagonism—on the most friendly basis—but we are bound to win, gentlemen, if we get your support as De Beers shareholders.' And, of course, win they did. The take-over of these two mines was speeded up by the disastrous reef falls that they suffered in 1888. Rhodes had prophesied that this would happen and was ready to act when it did. He was lunching at his club when one of the important owners of Bulfontein arrived to announce the disaster. A cab was called and Rhodes, accompanied by a few friends, rushed to view the scene. After sitting for an hour on the edge of the mine, oblivious of the crowd of sightseers, Rhodes presented his proposal for amalgamation. The owners of the two mines had no option but to commence negotiations. These negotiations were protracted (some companies held out for years) but De Beers Consolidated Mines triumphed in the end. It was always accepted that they would.

[3]

Not everyone welcomed the advent of De Beers Consolidated Mines. The diamond industry, as a whole, benefited by amalgamation but many diggers and small company owners suffered. As a result of the reorganisation of the mines, a large number of employees were thrown out of work. Attempts were made to find suitable alternative employment, but it was impossible to accommodate all the unemployed. Resentment against Rhodes rose to such a pitch that, for a time, he had to be given police protection whenever he was in Kimberley. The widespread ill-feeling sparked off at this time was to pursue Rhodes throughout his life. Not every family in Kimberley was able to accept him as the hero of Empire that he later became.

Nor did Barney Barnato escape the hostile demonstrations that followed amalgamation. In November 1888, he stood for election to the Cape Assembly. His election campaign was one of the most flamboyant ever conducted in Kimberley. He toured the streets in a carriage decorated in gilt, drawn by four silver-harnessed horses and attended by outriders wearing jockey caps and breeches. Two young boys, dressed in olive green velvet coats and sporting cockaded hats, accompanied him wherever he went. He was elected with a large majority, but not without a tough fight. At almost every meeting he addressed he was heckled mercilessly, bottles were thrown, and the proceedings often ended in a free fight. Inevitably, the old taunts of I.D.B. were resurrected. 'Men are being put forward for election', commented J. X. Merriman in Cape Town, 'who, if returned, would be a disgrace to any society, and it is quite possible that we may see the spectacle of the dupe on the Breakwater and his employer in Parliament.'

Barney was never popular as a Member of the Legislative Assembly. The Kimberley papers were constantly attacking him for neglecting his duties. In a political demonstration in 1891, both he and Rhodes were burnt in effigy by an angry Kimberley mob. But, by then, Barney Barnato was very different from the youngster who had delighted audiences in the old Theatre Royal. He no longer had to bow to the gallery. He was rich enough to be cast as a villain, even though he preferred to play the clown.

There was one last attempt to prevent amalgamation. It almost succeeded. In August 1888, a group of Central shareholders challenged the decision to merge their company with De Beers.

They argued that, under their deed of association, such a merger could only be effected with a 'similar company'. The Trust Deed drawn up by Rhodes showed that De Beers Consolidated Mines, whatever else it might be, was vastly different from the Central Company. They took their case to the Cape Supreme Court.

Their counsel had no difficulty in proving their point. He argued that the sole business of the Central Company was 'to dig for diamonds in the Kimberley Mine'. The new corporation, however, was empowered to do anything and everything. 'Since the time of the East India Company, no company has had such power as this,' he declared. 'They are not confined to Africa: they are authorised to take steps for the good government of any country . . . they could annex a portion of territory in Central Africa, raise and maintain a standing army, and undertake warlike operations.' What, he wanted to know, had all this to do with digging for diamonds?

The Court agreed. It was clear that diamond mining in Kimberley formed only an insignificant part of the powers acquired by De Beers Consolidated Mines. The company, the judge observed, was free to mine anywhere in the world. It could mine diamonds, gold or coal; it could carry on banking operations and undertake financial obligations for foreign governments; it could even become a water company. 'The powers of the Company', he said, 'are as extensive as those of any Company that ever existed.' He could not do otherwise than give judgement in favour of the plaintiffs.

Rhodes was not unduly worried by the verdict. 'If you cannot manage a thing one way, try another,' he was found of saying. He quickly found another way to take over the Central Company. The judge, in his summing up, had hinted at a possible solution; Rhodes acted upon it. He and Barnato held the majority of shares in the Central. They now used this majority to put the company into voluntary liquidation. The shareholders opposed to amalgamation could do nothing. The total capital of the Central Company was £1,779,650 and of this De Beers Consolidated Mines held shares representing £1,660,930. The dissidents were very much a minority.

The Central Company was put into liquidation on 29 January 1889. In the course of liquidation the Kimberley Mine had to be tendered for. The highest tender came from De Beers Consolidated Mines. On 18 July 1889, De Beers issued a cheque in payment for the purchase of the Kimberley Mine. The cheque became world famous; for years replicas of it were sold in Kimberley as souvenirs.

Made out in favour of the liquidators of the Kimberley Central Diamond Mining Company Limited, it authorised payment of £5,338,650.

The passing of this cheque meant more than the extinction of the Central Company; it marked the end of an era. After almost twenty turbulent years, diamond mining in Kimberley ceased to be an adventure and became a prosaic, soulless industry.

Not only mining, but Kimberley itself changed. The lively competitive spirit which had distinguished it from other South African towns gradually disappeared. Company mining had long since sapped much of the town's vitality; now, dominated by one vast corporation, it lost its individuality. No longer was it a place where fortunes could be made or lost overnight. No longer did it attract the speculators, the gamblers and the rogues who, whatever their motives, had brought a whiff of romance to the arid wastes of Griqualand West. Many of the characters who had made Kimberley what it was had, it is true, left the town already; amalgamation speeded the last of them on their way. They were replaced by the efficient, responsible employees of De Beers Consolidated Mines.

'This town,' noted a visitor a few years later, 'once humming with speculation, business, and movement [is] now the essence of a sleepy respectability and visible prosperity. . . . The stranger soon perceives that the whole community revolves on one axis, and is centred, so to speak, in one authority. "De Beers" is the moving spirit, the generous employer, and the universal benefactor.' This is what many of the citizens of Kimberley had always hoped for. Whether they welcomed it when it came is another matter.

PART THREE
THE MAGNATES

CHAPTER EIGHTEEN

ROBINSON ON THE RAND

FOR years men had searched for gold in South Africa. Long before diamonds were discovered prospectors had been roaming the veld, panning gravel from river-beds and collecting rock samples in the hopes of finding gold. Gold-strikes and rumours of gold-strikes had been reported, at one time or another, from practically every part of the sub-continent. Such reports rarely stood up to close examination. More often than not they proved nothing more than wild exaggerations; sometimes they were found to be completely false. But men went on looking.

They looked mostly in the Transvaal. Here, in the tawny, rock-studded high-veld, it seemed fairly certain that large quantities of gold would one day be discovered. The signs were promising. A lone prospector had discovered gold-bearing quartz there as far back as 1853. The Boer government, anxious to preserve their tranquil way of life, had sworn him to silence, but the news had leaked out. There were reports of other finds. Then, in 1871—the year of the great diamond rushes—an alluvial gold-field was dis-covered in the eastern Transvaal. News of this discovery had created considerable excitement at the diamond diggings, but the reports were not sufficiently tempting to lure diggers from Griqualand West. Not until three years later was there a rush to the gold-fields of the eastern Transvaal. Many men, including Cecil Rhodes's brother, Herbert, left Kimberley but most of them were quick to return. Panning for gold proved less rewarding than digging for diamonds. Among the first arrivals, only a few, made any real money from these alluvial diggings.

Ten years later, in 1884, payable gold was again discovered in the Transvaal; this time, further to the east, not far from the Portuguese border. Here, in De Kaap Valley, the signs were promising indeed. The little town of Barberton, which developed as more and more diggers were attracted to the eastern Transvaal, looked at one time like becoming a second Kimberley. Beit and Rhodes displayed keen interest and many diamond diggers sold up and went off in search for gold. There was a boom in gold shares in Britain. But the gold finds —although often incredibly rich—proved to be limited and Barberton was destined to decline almost as rapidly as it had risen.

Interest in Barberton was still strong in the middle of 1886 when the real gold riches of the Transvaal were revealed. A gold-bearing reef was stumbled upon some thirty miles from Pretoria, the capital of the Transvaal, in a district known to the Boers as the Witwatersrand—the Ridge of White Waters. The reef uncovered on the Witwatersrand—or the Rand, as it came to be called—was later found to stretch for thirty miles from east to west. Its discovery marked the beginning of South Africa's great gold-mining industry. The town of Johannesburg, upon which it was centred, soon rivalled, then eclipsed, Kimberley. But it was the diamond magnates who supplied the money, and much of the knowledge, which developed South Africa's gold mines.

[2]

One of the first men in Kimberley to hear of the fabulous gold finds on the Witwatersrand was J. B. Robinson. He was often to relate how the news first reached him.

Sitting one day in the vestibule of the Kimberley Club, he was handed a telegram. It came from a diamond digger whom he had once helped. It read: 'A discovery has been made about 30 miles from Pretoria, of a peculiar kind of conglomerate, shedding gold freely, I think it is worth your while to come on at once and see it.' While pondering whether or not to follow this up, a mining expert walked into the club; Robinson showed him the telegram. 'Pooh,' said the expert, 'it is all bosh; it's a river wash tilted up, and will pinch out.' But, as Robinson had never taken much notice of experts, he decided to ignore this one. He made for the coach office where, after some difficulty, he managed to book a seat on a coach going to the Transvaal. He left Kimberley on a Sunday and went on to make his second fortune.

That is how Robinson told it. The truth must have been somewhat different. His decision to go the Transvaal was not arrived at so coolly. It was the desperate gamble of a man on the verge of bankruptcy.

By the middle of 1886, J. B. Robinson's fortunes were at their lowest ebb. The first half of that year was one of the most disastrous periods in his life. For it was in January 1886 that he had been forced to confess his dire financial position to J. X. Merriman and abandon all hope of playing a part in the amalgamation of the diamond mines. Then, to add to his humiliation, he suffered a serious political defeat. A by-election was about to take place. The only candidate in the field was a man known to oppose Robinson's great source of pride—the Diamond Trade Act. In an attempt to bolster his declining fortunes, Robinson decided to enter the contest. He fought hard; but the campaign only brought him further misery. While he was electioneering his youngest daughter, aged four months, died. Despite his obvious grief, he continued to address meetings. His fight was as much for the remnants of his reputation as for a seat in the Cape Assembly. His efforts were in vain. When the results were published on 7 May, his opponent had gained a clear victory.

Robinson took his defeat badly. As truculent as ever, he petitioned to have the election declared invalid. He accused his opponent of not being a citizen of the Cape Colony and of employing corrupt practices. It was while he was awaiting the result of this petition that he received news of the gold finds in the Transvaal.

The news could not have come at a more inauspicious moment. The Cape of Good Hope Bank was threatening to sell the last of his diamond shares. Over the past few months, he had tried to obtain help from every financier in Kimberley. They had all refused him. Now, acting upon advice, he turned to Alfred Beit. It must have hurt his pride to do so. Beit was in league with Rhodes and Robinson hated Rhodes. Nevertheless, Beit was his last hope.

According to Beit, Robinson did not ask for help, he demanded it. He burst into Beit's office and accused Beit and Rhodes of ruining him. He made it appear that Beit was under obligation to rescue him. Anyone other than Beit would have sent him packing. But soft-hearted 'little Alfred' gave in to him. He had no need to. In no way was he responsible for Robinson's plight, but he could never turn a deaf ear to anyone who appealed to him for help.

'Beit arranged to stay execution,' says Sir Percy Fitzpatrick in a private letter, 'and finally arranged to take over Robinson's position from the Cape of Good Hope's Bank. It was then that Robinson bewailed and bemoaned his dreadful bad luck because he had infor-mation about gold discoveries in the Transvaal, and had a "hunch" that he was going to make by far the biggest fortune he had ever

made. Eventually he proposed that Beit should finance him and send
him up to the Transvaal to acquire some gold properties there. As
well as saving him [in] the Cape of Good Hope Bank matter, Beit
financed Robinson both for capital to be expended, and for main-
tenance, travelling and living; in fact he sent him up as representing
his firm or syndicate in the usual way.'

Robinson's debt to the Cape of Good Hope Bank appears to have
been larger than he originally admitted to J. X. Merriman. One
estimate puts it at £80,000. And Beit himself was never quite sure
how much he advanced Robinson for expenditure in the Transvaal.
He thought it might have been £20,000 or £25,000 (others have put
it at £30,000). But whatever sums of money were involved, they were
undoubtedly considerable. So was the risk taken by Beit. Yet Beit
made no mention of the transaction until he told Sir Percy Fitzpatrick about it several years later. He then treated it as a joke. 'Beit
told me in his usual good-natured, laughing way,' wrote Sir Percy,
'all about how Robinson came to go to the Transvaal. Beit, as usual,
only laughed at the bombastic, self-advertisement of J.B.R. yet the
revelation was quite new to me.'

One of J. B. Robinson's grand claims was that he had pioneered
the Transvaal gold fields. He never tired of boasting of his enterprise,
his far sight and his success. But there is no record that he ever
acknowledged the debt he owed Alfred Beit. It is not surprising that
Beit's friends came to dislike Robinson intensely.

[3]

There were several groups of diggers at work when Robinson arrived
on the Witwatersrand. He made a rapid tour of the diggings and
then stopped for a closer inspection. Going to a spot where a family
named Bantjes had dug to a depth of about 12 feet, he took a few
samplings and put them in his white pith helmet. Any doubts he
may have had were quickly dispelled. 'I took some stuff,' he was to
say, 'and panned it in a stream of water. The moment I saw the gold
in the dish and could see the reef running in the way it did, I
decided that a grand discovery had been made.'

He lost no time in acting on his belief. Inspanning the mule cart
he had hired, he drove along the reef until he came to a farm called
Langlaagte. Here, helped by his knowledge of the Boers' language, he
was able to buy part of the property from the widow Oosthuizen. The

negotiations took three days and, by all accounts, Robinson drove an extremely hard bargain. He secured the property for £6,000. This purchase was to become famous in gold mining history. For with his uncanny instinct, Robinson had driven straight to the heart of the Witwatersrand. Langlaagte was to prove one of the most valuable properties on the Transvaal gold fields. No mining expert, armed with technical data, could have selected a better mining proposition.

Shortly after this he bought a half share in the claims of a digger named Japie de Villiers. This buy was equally inspired. Later, when the other half of the claims had been acquired, a mine—named after Robinson—was to produce ore which yielded as much as ten ounces to the ton from these claims. If for nothing else, Robinson deserves full credit for the genius which led him to make these purchases. Without any geological training, he appears to have recognised the nature of the reef and to have bought his properties accordingly. This, more than his boasting, entitled him to rank among the important pioneers of the Witwatersrand. The same genius led him to purchase what were to become his Randfontein estates.

He went on a buying spree. Others laughed at him and the experts taunted him with snatching up 'cabbage patches'. 'This served me very well,' he said, 'as I went on buying and people threw their claims at me for something like £50. It suited my book admirably, because the more the fields were denounced the more properties I bought.' Not all the farmers he approached were so eager to sell. Once other capitalists began to arrive from Kimberley competition set in. Many outlandish stories were told about the lengths to which Robinson would go to outwit his rivals. Although most of these accounts are suspect, one became so popular that it is worth retelling. Alfred Beit was involved in the deal.

Beit visited the Rand to find out how Robinson was faring with their syndicate. He was, of course, delighted with the purchases that Robinson had made. Robinson then told him of a particularly valuable farm in which he was interested. It was owned by an old Boer couple; the price asked for it was £30,000. As Robinson spoke the Boers' language fluently he insisted on carrying out the negotiations personally and on this occasion asked Beit to accompany him. Somewhat reluctantly Beit agreed.

They drove to the farm and the lengthy negotiations began. The farmer and his wife, suspicious of their smart visitors, were reluctant to commit themselves. Late into the night they discussed the sale

until it was eventually decided that Robinson and Beit would sleep
at the farm. There were no spare beds and, to Beit's horror, the two
magnates were obliged to doss down on animal skins in the living
room. Beit was even more alarmed the following morning when he
found that he was expected to drink endless cups of weak, sugarless
coffee. He could not swallow the stuff. Robinson, however, realised
that to refuse would be to offend. He gallantly sipped at his own cup
and, immediately they were alone, downed Beit's as well. Each time
the farmer's wife returned he complimented her on the coffee and
asked for more.

His tactics appeared to work. The lady, flattered, showed signs of
weakening. Then, unexpectedly, she burst into tears. After some
coaxing, Robinson discovered the cause. The sale was to be an
outright one and she despaired of parting with a particular treasure.
To the astonishment of the financiers, she explained that this was a
miserable pink geranium, planted in a tin on the window sill. This
plant had been given to her by her dead daughter; she felt that she
could not honour the sale by parting with it. Beit, touched by her
simplicity, was eager to reassure her. Not so Robinson. He made a
great show of wanting the geranium. When he finally relented, with
seeming reluctance, the woman was so relieved that she summoned
her husband and concluded the sale before he could change his mind
again.

There are several versions of this story. Beit told this one to Sir
Percy Fitzpatrick. Legend has it, however, that Rhodes was also
involved in this deal and that while Robinson was reducing the
farmer's wife to tears, Rhodes was outside trying to negotiate a sale
with her husband. It seems unlikely.

Robinson was not always so successful in his deals, but his investments were valuable enough to compensate him several times over
for his Kimberley losses. His purchase of the Randfontein estates
was to make him a multi-millionaire. Within a matter of weeks he
had risen from the verge of bankruptcy and laid the foundations of a
second, much larger, fortune. It was an astonishing achievement.

Surprisingly, it was an achievement he was never to acknowledge.
His reluctance to admit to his greatest triumphs is, perhaps, the
strangest facet of his puzzling personality. If it is true that he was
bankrupt when he arrived at the river diggings, his success in
Kimberley was remarkable. He had undoubtedly lost his Kimberley
fortune when he arrived on the Rand and this makes his second

success astounding. Yet he steadfastly refused to acknowledge his impoverishment and consequently his real claim to fame has been overlooked. Was this concealment the result of some quirk of pride? Or was it, as his enemies hinted, that his methods of making a fortune did not bear close examination?

Whatever the truth, once he was back in the saddle there was no shifting him. In 1887, when Rhodes and Beit took time off from their Kimberley struggle and arrived at Pretoria to attempt an amalgamation of the gold mines, Robinson refused to cooperate. Rhodes had usurped his position in Kimberley and he had no intention of allowing the same thing to happen on the Witwatersrand. The feud between the two of them intensified.

Robinson had no option but to work with Alfred Beit. The gold prospecting syndicate which Beit had financed—known as the 'Robinson Syndicate'—had been divided three ways. Robinson and Beit each had a third interest and the other third was held by Robinson's old partner, Maurice Marcus. Robinson appears to have acted for himself and Marcus while Beit appointed two young men, Hermann Eckstein and J. B. Taylor, to represent him in Johannesburg. It was not long before Robinson was at loggerheads with Beit's representatives. Beit had foreseen what would happen. 'He made liberal terms,' says Sir Percy Fitzpatrick, 'but he knew that Robinson was a man whom it was impossible to work with and that a break would have to come.'

The break came, much to the relief of Eckstein and Taylor, in 1888. It came at Robinson's suggestion and, as it worked out, to his disadvantage. Having displayed his genius by buying the Langlaagte properties, he and Marcus agreed to sell a portion of these concerns to the firm of Wernher and Beit. After lengthy negotiations, Robinson and Marcus were bought out for £250,000. It was one of the most unfortunate deals they ever made. For, it is estimated that Wernher and Beit eventually reaped something like £100,000,000 from their share of the Robinson Syndicate.

However, when Robinson sold he was convinced that he had made a splendid bargain. Both he and Marcus were bowled over at receiving a cheque for £250,000. They told J. B. Taylor that they had never seen a cheque for so large an amount in their lives. The sight of it moved Robinson to an unusual display of generosity. He invited Eckstein and Taylor to join him and Marcus for a glass of champagne. They went to a bar and Robinson expansively ordered four glasses of wine. As they were about to raise their glasses, Abe Bailey, another Rand pioneer, walked into the bar. Robinson spotted him.

'Come along Abe,' he shouted, 'we are celebrating a big share transaction that I have just put through with Eckstein and Taylor. Come and join us.'

Then he reached across, took Marcus's glass, and handed it to Bailey.

'Marcus, you don't want any,' he said.

[4]

Robinson was no more popular in Johannesburg than he had been in Kimberley. He never identified himself with the mining community. As always his eyes were turned to the seats of power. In Kimberley, under British rule, he had posed as an ardent imperialist. He had been president of the Kimberley branch of the Empire League and had emphasised his British connections. Such allegiance was of doubtful advantage in the Transvaal. He was quick to play it down.

Almost from the moment he arrived on the Witwatersrand, Robinson set out to cultivate the Boer authorities. He had good connections. His elder brother, William, was a Transvaal burger and had once been a candidate for the Presidency of the Boer republic. Robinson made a great show of speaking the language of the Boers and claimed President Kruger as an intimate friend. 'I am on very friendly terms with the old man,' he would say. This appears to have been true. At Robinson's persuasion the suspicious old President agreed to descend the shaft of a gold mine. Few Rand capitalists could have won Kruger's trust to that extent. However, the friendship did nothing to increase Robinson's popularity with the miners of Johannesburg. There was little lost love between President Kruger and the Transvaal's mining community.

The Boers were wary of the crowds of foreigners—or uitlanders, as they called them—who had invaded their country. They feared that this, largely British, community would one day rob them of their hard-won independence. Paul Kruger, the stubborn, patriarchal President of the Transvaal, was determined that this should not happen. When he was a child his parents had joined the large parties of Boers who had trekked from the Cape to escape the British; he intended to preserve his people's freedom. In his efforts to keep the uitlanders at bay Kruger was unrelenting. He refused to allow them any say in the government of the country. He taxed them heavily and hemmed them in with petty restrictions. So firm was his grip that it soon became a stranglehold.

The uitlanders responded with heated protests. They accused Kruger of stifling their enterprise and ruining their industry. They claimed their rights as taxpayers: they demanded the vote. Soon it became a matter of supporting one side or the other. Robinson's support for Kruger placed him, as far as the uitlanders were concerned, squarely in the enemy camp.

This did not worry him. He could afford to dissociate himself from the struggles of the uitlanders. He was not particularly concerned with the internal politics of the Transvaal. As long as he could make money, he was prepared to pander to the Boer authorities. Johannesburg, for him, was simply a place of business; he never became personally involved in the daily life of the town. His days of political in-fighting were over.

Once he had established his gold mining interests, Robinson left South Africa. He moved his family to London and took his place among the growing throng of South African *nouveaux riches*. He was one of the richest. Dudley House in Park Lane became his home. His wife set herself up as a musical hostess, attracting performers such as Melba, Réjane and Clara Butt to her concerts. He sent his sons to Eton. He bought a yacht, he collected paintings, he entertained royalty. He shaved off his beard and even exchanged his pith helmet for a top hat. He did, in fact, everything that was expected of a parvenu millionaire. Only in the unaltered steeliness of his glance and the tight set of his mouth could one recognise the ambitious young man who had once ridden through the dust of the Free State veld, scratching a living as a trader.

Not that he was ashamed of his origins. Like many a self-made man, he delighted in drawing attention to the contrast between his past and his present. 'I remember a concert given by J. B. Robinson in Dudley House,' says Lady Glover, '—a wonderful display of orchids and fairy lights, every delicacy in and out of season and the choicest wine in profusion, and the divine Sarah among the most prominent artistes of the day. For all the wealth and display in his entertainment, Mr Robinson seemed a simple man in his tastes. A fashionable lady is said to have apologised to him because she could not give in her house accommodation worthy of his magnificence. "And to think", he chuckled, "that I remember the time when I was glad to sleep on the ground under a tent."'

CHAPTER NINETEEN

MAN OVERBOARD

BARNEY BARNATO was late arriving on the Rand. While other Kimberley capitalists were scrambling for gold claims, he was tied down by his struggle for control of the diamond mines. Not that he thought he was missing much. For a long time he remained convinced that there was more to be made out of diamonds than out of gold. He had good reason for his belief. Gold rushes in the Transvaal were nothing new. He had burnt his fingers slightly by investing in some of the earlier gold companies that had collapsed; he had no wish to repeat the experience. What he heard about the Witwatersrand was not particularly encouraging. His own mining experts were sceptical and, at first, he was prepared to accept their judgement.

At the beginning of 1888, he paid a brief visit to Johannesburg. A lightning tour of the claims confirmed his doubts. 'I will confess,' he later admitted, 'that I did not then form a bright opinion of the goldfields.' But the gold fever continued and, eventually, he was infected. Some ten months after his first visit, he returned to the Witwatersrand. This was at the end of 1888. The amalgamation of the diamond mines was nearing completion and he had recently won his election to the Cape Assembly. Now he was able to take time off to make a closer study of developments in the Transvaal. He was immensely impressed.

On this, his second visit, he was given a great welcome. Fresh from his Kimberley triumphs, he was very much of a hero. Barnato was now a powerful name in mining circles: everyone was anxious to hear what he had to say. He did not disappoint. Shortly after his arrival he spoke at a banquet given in his honour. The inspired phrase with which he closed his speech was to rank among his more memorable utterances. 'I came for a visit,' he declared, 'but I shall stay for months, and I look forward to Johannesburg becoming *the financial Gibraltar of South Africa*.'

He lost no time in demonstrating his confidence. For the next few weeks, he went about buying up properties with an enthusiasm which startled even those who had known him in Kimberley. Hermann Eckstein was soon reporting to Alfred Beit on the great show Barney was making in Johannesburg. The Barnatos, he said, were intent on edging in on any properties that the Beit concern had not already bought. It was obvious that the Kimberley alliance was not to be continued on the Rand. 'Beit don't like me up here,' said Barney, shortly after his arrival, 'don't care, don't care, I know 'em.' Once again Barney Barnato was a rival to be contended with.

He got off to a spectacular start. Within three months, it is estimated, he had invested some £2,000,000 in property and mining shares. He established several gold mining companies—including the New Primrose, the New Croesus and the Glencairn Main Reef —formed a large estate company, planned the Barnato Buildings in Commissioner Street and took over the Johannesburg Waterworks. Woolf and Solly Joel were there to help him, but it was Barney who made the decisions. He held long consultations with his mining engineers, inspected mines, closed property deals and interviewed prospective employees. Nothing was too trivial to command his attention. 'I must look into everything that concerns me for myself,' he told a reporter.

So hard did he drive himself that, in February 1889, he collapsed. He became so seriously ill that his nephews decided to call a doctor. They sent for Dr Mathews, who had moved from Kimberley to Johannesburg. Mathews was pessimistic; he warned Barney that if he continued at his present rate he would kill himself. Barney was amused. 'What a devil of a fight there will be over the chips,' he laughed.

Once he had recovered, Barney plunged back into a whirl of activity. There was no question of his letting up. Despite the demands made upon him, he was in his element. In many ways, early Johannesburg was like early Kimberley. It was a rough, dusty, primitive, tin-roofed mining camp: a place where men drank, gambled and fought, and where only the toughest survived. Many of Barney's disreputable old cronies were there. He met them in the streets, in the bars and on the race course. He was happier with them than he could ever be with men like Cecil Rhodes and Alfred Beit. Soon after his arrival, he bumped into his former partner Lou Cohen. They went for a drink and Barney boasted of the part he had played in the amalgamation of the diamond mines. He was still convinced that he had got the better of his opponents.

'You see I beat the swines down below,' he told Cohen, 'beat 'em hands down—nobody didn't want me in, not even Rhodes, though he was on my side—he had to—he had to. But I know he doesn't like me—he looks down on me because I have no education—never been to college like him—ah, ah, but I had my way—his education! —he had a stone in hand, but I won in a canter.'

This was how Barney Barnato liked to see himself. He was the poor boy who had made good: a natural genius who had learned his lessons from life, not books. While this was true enough, it only partly explained his success. His was a far more complex personality. The image he projected was largely of his own making. A sharp-witted cockney from Whitechapel, he knew what the public expected and he saw that they got it. He presented himself as vulgar and shrewd; cocky and big-hearted; a dandy and a joker. So complete was the caricature that neither his friends nor his opponents saw through it. In the role of Barney Barnato, the self-made millionaire, he gave his greatest performance. He succeeded in fooling everyone except himself.

He went on buying properties, forming companies and making money. Not all his ventures were successful. There were times when his ill-informed enthusiasm got the better of him. Experts had a poor opinion of him as a mining man. 'He had no conception of the means by which an 8-dwt property can be made to pay,' it was said. 'Further, he took a tradesmanlike view of finance, and always pushed his wares to the highest figure . . . there was a lack of repose in his policy. He must always be amalgamating, or reconstructing, or altering fundamental bases, not altogether to his own advantage, to say nothing of the advantages of the shareholders.' His restlessness reflected his basic insecurity. He worried too much. He could never let well enough alone. He lived on his nerves. Some of his financial experiments brought ruin to his fellow shareholders. Although he was often astute enough to withdraw from a faltering company before the crash came, others suffered. Fairly or unfairly, he was blamed for several lost fortunes on the Witwatersrand.

He did not confine himself to money making. The brash, carefree atmosphere of Johannesburg's camp life rekindled the passions of his early Kimberley days. In the Diamond City he had made a fortune but lost much of his popularity: Johannesburg gave him the chance to start again. Once more he became a familiar figure in the bars and billiard rooms. He arranged sparring matches, promoted prize fights and occasionally appeared in some of his well-known roles at the local theatre. Whatever he did, he was sure of a wide audience. For all his riches he was reluctant to lose touch with the public. It was not easy for him to recapture his youth. He was in his late thirties and the strain told on him. He became moody, short-tempered and began to drink heavily. Emotionally he was ill-equipped to bridge the gap between early Kimberley and early Johannesburg.

His impact on the Rand was spectacular but brief. It lasted little more than seven years. He made a great splash on the Transvaal gold fields, but he was never an important factor in the development of the gold mining industry. Once he had established his interests in Johannesburg, he spent less and less time in South Africa. It was his nephew, Solly Joel, who was largely responsible for the Barnatos' progress on the Witwatersrand.

[2]

In July 1892, Fanny Bees fell pregnant. She and Barney had lived together for fifteen or sixteen years without producing children. If Barney had ever considered marrying her during that time, there had been no pressing need to make their union legal. Fanny had been content to remain in the background and they had both, it seems, been reluctant to defy openly South African social conventions. Now, with a child on the way, Barney decided to make Fanny his wife. He was rich and powerful enough to ignore outside prejudices; above all he was determined to acknowledge his own child. Much as he doted on his nephews, he had always longed for a son. There could be no question of his foregoing his rights as a father.

Barney and Fanny were married at the Strand Register Office in London on 19 November 1892. Their first child, a daughter, was born on 16 March 1893. She was named Leah Primrose, after Barney's mother and the first Barnato gold mining company. The longed-for son was born in March the following year. Barney toyed with the idea of calling this boy Ladas, after the Derby winner of that year, but finally settled for Isaac Henry. Like his cousin, Isaac Joel, young Isaac Barnato was always known as Jack. A second son, Woolf Joel ('Babe') born in September 1895, completed the family.

For each confinement, Barney took his wife to London. 'I take no risks,' he said. And during these years he contemplated making his home in England. He decided to build a house for himself in Park Lane, within walking distance of J. B. Robinson. The site for this house cost £70,000 and Barney planned to build it on a lavish scale. It was to be a five-storeyed mansion, with a marble staircase, an enormous ballroom, two billiard rooms, a schoolroom, and heated throughout by radiators. Built in a florid mixture of styles, it became, when it was completed, a garish symbol of South African opulence. 'Everybody who passes down Park Lane', it was said, 'was reminded, by a certain house sprawling with naked nymphs and cupids, that the shortest way from Whitechapel to Mayfair crossed and recrossed the Equator.'

Barney took great pride in the building of his house. He rented Spencer House off St James's Street as a temporary home and kept a keen eye on progress in Park Lane. For him the house was both a hobby and an advertisement. After years of dusty discomfort in Kimberley, he intended to live in luxury and to dazzle the world with his entertainments. He bored his friends by taking them on inspection tours of the building but, if Lou Cohen is to be believed, was somewhat doubtful of his chances of becoming a social success. 'I did get the "sicks" over that Park Lane palace,' Cohen says. 'Every other day Barney would insist on my going over it until I hated the building. I asked him if he intended to entertain largely, and he replied in dubious tone, "Yes, if they'll come." '

But if not everyone in London accepted him, it was difficult to ignore him. Everything he did became news. The gossip columns were full of stories about his house, his horses and his eccentricities. He was far and away the most colourful, the most amusing, the most controversial of the South African *nouveaux riches*. Suburban tea parties were enlivened by stories of his outlandish behaviour. His audacity was the refrain of a music-hall song:

> 'I'm beautiful, bountiful Barney
> And Beit may go to pot!
> Beautiful, bountiful Barney
> And Robinson may trot!
> The public all run after me
> For I know how to blarney.
> I tell you straight, he's up-to-date,
> Is beautiful, bountiful Barney!'

His greatest triumph as a London celebrity came on 7 November 1895, when the Lord Mayor, Sir Joseph Renals, gave a banquet in his honour. Unfortunately, the occasion was somewhat marred by the widespread bad feeling it aroused. In his introductory speech, the Lord Mayor admitted that he had received a flood of anonymous insults for honouring Barney with a dinner. And Barney did not help matters by larding his own speech with quotations from the Stock Exchange. What should have been a dignified civic function looked, to many people, like a cheap advertisement for the Barnato Brothers' companies. 'The Mansion House', commented the *Times*, the next day, 'is not the proper place for glorifying a successful operator of the Stock Exchange.' Others papers were still more scathing.

Barney had good reason for trying to boost his own enterprises. By November 1895 some of his financial ventures were running into trouble. The most important of these, the recently founded Barnato Bank, Mining and Estates Company, was particularly shaky. Launched with a capital of £2,500,000 it had sparked off a typical 'Barnato boom'. Shares in it were so eagerly sought that they doubled and trebled in value overnight. But the time was not right for such speculation. A panic on the Johannesburg Stock Exchange boomeranged in October 1895 and hit the Barnatos hard. Barnato Bank shares plummeted. No reports or accounts of the new company were available and disillusioned shareholders began to demand their money back. Once again the Barnatos were accused of misleading the public and ruining unsuspecting investors.

Barney's efforts to restore confidence in his firm were in vain. What little chance they had of success was completely ruined at the beginning of 1896, when the Witwatersrand was hit by the most serious crisis it had yet experienced.

[3]

By the end of 1895 the long smouldering discontent among the uitlanders of the Transvaal was threatening to flare into open rebellion. Repeated attempts by the uitlanders to reach a political settlement with the Kruger regime had all failed. There were faults on both sides, but the mining community was driven to the conclusion that unless they could force the Transvaal government to allow them some say in the running of the country there could be no hope for the gold mining industry.

One of the principal driving forces behind these insurrectionary sentiments was Cecil Rhodes. For his own imperialist reasons, Rhodes was anxious to topple the Kruger government; he seized upon the inflammable situation in the Transvaal to further his own ends. A plot was hatched. Rhodes, with the help of Alfred Beit, agreed to finance the arming of the uitlanders who were to rise and demand their political rights at gun point. They were to be assisted by a force of professional soldiers under the command of Rhodes's Kimberley friend, Dr Leander Starr Jameson. Arrangements were made for Dr Jameson to assemble his troops in Bechuanaland, close to the border of the Transvaal. When Johannesburg rose in rebellion, Jameson and his men were to invade the Transvaal on the pretence of coming to the 'rescue' of their fellow countrymen. It was the failure of this immaturely conceived rebellion that dealt a double blow to Barnato's efforts to salvage his bank.

For fail it did. There was little or no co-ordination between Dr Jameson and the conspirators in Johannesburg. At the last minute the plot went haywire. The Reform Committee—which was directing the Johannesburg side of the affair—discovered a great many weaknesses in their organisation. They tried to delay the rising and sent messages to Jameson warning him not to come. But they left it too late. Jameson refused to be put off. Suspecting the Reform Committee of cowardice, he decided to force the issue. On the appointed day, he rode across the Transvaal border and led his men into an ambush. The Boers had got wind of the conspiracy and were waiting for him. Johannesburg failed to rise and Jameson was forced to surrender. A few days after Jameson's capture (on 2 January, 1896) the members of the Reform Committee were rounded up and sent to gaol in Pretoria. One of the last to be arrested was Barney Barnato's youngest nephew, Solly Joel.

The trial of the Reformers was expected to take place in April 1896. At the beginning of March Barney and Fanny, wearing matching fur coats, sailed for South Africa. As always when one of his nephews was in trouble, Barney was in a highly emotional state. He went straight to the Transvaal and did his best to console the imprisoned Solly Joel.

Throughout the trial, which was held in the Pretoria Market Hall, Barney was seen to be on edge. He had no faith in the court proceedings and considered the Reformers' decision to plead guilty 'an act of suicidal mania'. His conviction that they were throwing their chances away was confirmed when sentence was passed. Four leading members of the Reform Committee were sentenced to death and the rest to two years' imprisonment and a fine of £2,000.

Solly Joel was among those who received the lesser sentence, but this in no way placated Barney. Rushing into the street, he gave way to one of his neurotic outbursts. He roundly abused Kruger and swore that he would stop at nothing to get the prisoners released. Later, meeting the man who had passed the severe sentences, Judge Gregorowski, he treated him to every insult in his considerable vocabulary. 'Mr Barnato,' gasped the judge, 'you are no gentleman.' 'And you are no judge, Mr Gregorowski,' Barney shouted back.

Barney's behaviour immediately after the trial bordered on the eccentric. Dressing himself in black, he bound his hat with crape and advertised that 'all our landed properties in this State will be sold by public auction on Monday, May 18, 1896'. He adjourned his company meetings and let it be known that an expenditure of £200,000 a month from his mines would be lost to the Transvaal as a result of his decision. He could offer no greater threat to the Transvaal authorities. It had some effect. President Kruger agreed to an interview with the funereal, crape-crowned Barney and assured him that he was doing his utmost to obtain a mitigation of the sentences.

As it happened, the sentences were quickly reduced. Within twenty-four hours of the judgement the death sentences were commuted and a few weeks later the four leading Reformers were set free on payment of £25,000 each. This money was paid for them by Rhodes and Beit. The other members of the Reform Committee had obtained their release eleven days earlier, on 30 May. Each of them was fined £2,000. Imprisonment had been no hardship for Solly Joel. Not only was he supplied with champagne, caviar and roast duckling from a local hotel, but he had been able to round off his meals with the large Havana cigars that his wife smuggled into him in the crown of her hat.

How much the leniency of the Transvaal government was due to Barney's intervention is debatable. Kruger was being pressed from all sides to show mercy and, in any case, it is doubtful whether he meant the sentences to be anything more than a warning. However, Barney had no doubts about his success. 'No one else could have done what I have done,' he declared.

[4]

The strain of the trial and its aftermath had a decidedly adverse effect on Barney. He was never quite the same again. For a few weeks he seemed to brighten up, but soon his bouts of high spirits were interspersed with fits of black depression. He remained in South Africa until July. Business on the Rand had been seriously disrupted by the Jameson Raid fiasco and Barney and Solly Joel were hard pressed to put the affairs of their firm to rights. The responsibility oppressed Barney. 'I can't forget the work,' he told a friend. 'It's too much now. I feel it and yet I can't leave off.' He haunted the bars; drinking heavily and seeking reassurance from the sycophants who followed him around It took little to make him maudlin drunk. Those who had known him in earlier days became alarmed at the way he was deteriorating.

There was no improvement in his condition when he returned to England. Nor could he derive much comfort from the decline in South African gold shares on the Stock Exchange. Shortly after his arrival in London, the problem of the faltering Barnato Bank was solved when, without a word, it was merged into the umbrella Barnato concern—the Johannesburg Consolidated Investment Company. This questionable move again brought the Barnatos under fire. Shareholders who had had no say in the merger were loud in their protestations. The financial press deplored the transaction and hinted at sharp practice. Barney's life was threatened by anonymous writers who attacked him in poison pen letters.

The failure of the bank plunged Barney into deeper despair. He became restless, unable to concentrate; his memory began to fail. At night he would take long walks, armed with a stick, expecting all the time to be attacked by angry shareholders. His sleep was ruined by nightmares in which he saw himself penniless and pursued by creditors. Nothing could distract him. He lost all interest in the Park Lane mansion. He wandered about bleary-eyed and unshaven, unable to keep up the slightest social pretence.

Friends—including J. B. Robinson—urged him to take a holiday. In November he decided to act on their advice. On the spur of the moment he made up his mind to accompany Solly Joel, who was returning to South Africa. At first Solly was not keen on the idea; he managed to talk Barney out of it. However, just as Solly's ship was about to sail, Barney unexpectedly turned up at Southampton—

slightly drunk and wearing a pair of dress trousers and a sports coat under his fur-lined overcoat. He announced that he had come to see Solly off. After a few drinks, Barney became so maudlin that Solly took pity on him. Acting on an impulse, he invited Barney to join the ship and travel as far as Madeira. Barney leapt at the offer. He was still on board when the ship arrived at Cape Town.

In South Africa, Barney's behaviour became more erratic, more unpredictable. Arriving in Johannesburg he announced his intention of reorganising the Barnato group of mines from top to bottom; but it soon became obvious that he was not fit enough to undertake any real work. From time to time he went to the office and Solly tried to interest him in the firm's various activities, but to little effect. He was becoming more and more obsessed with the idea that he was on the verge of bankruptcy; the slightest fluctuation on the Stock Exchange sent him reeling to the nearest bar. Before long all Johannesburg was gossiping about his drinking bouts.

Early in 1897 he was joined by Fanny and the children. They brought Barney more worries. While the family was waiting to change trains at Vereeniging in the Transvaal, a railway official carelessly raised his rifle, which misfired, and shot Fanny Barnato in the leg. It was clearly an accident and Fanny was the first to forgive the official. But Barney magnified the incident and saw it as part of a plot. To his other obsessions he now added the conviction that the Boers were out to kill him and his family. His nightmares became worse. Waking in a sweat one night, he rushed from his house in his pyjamas and knocked up a neighbour, shrieking: 'They're after me. Let me in!'

On doctor's orders, he moved to the Cape at the end of April. He took Fanny and the children with him. The change of scene seemed to do him good. He went for daily drives and even took his seat again in the House of Assembly. But the improvement did not last. On 23 May, he suffered a relapse. In a delirium he started counting imaginary banknotes and, it is said, clawed at cracks in the wall looking for hidden diamonds. Fanny was terrified. She sent an urgent telegram to Solly Joel asking him to come to the Cape at once. Barney had partly recovered when Solly arrived, two days later, but the doctors advised him to keep his uncle free from business worries.

The following week the Barnatos sailed for England on the s.s. *Scot*. Barney had already booked a passage on the ship; now Fanny and Solly decided to accompany him and take the children with them. They left Cape Town on 2 June. Barney was dead before the ship reached Southampton.

During the first few days on board, he behaved quite normally. He chattered brightly with his fellow passengers, played card games and dominoes. Solly watched him closely to see that he did not drink too much, but felt no cause for alarm. Then the now familiar symptoms began to manifest themselves. Barney started fretting about the time the ship was making and went around asking the crew when they expected to arrive at Southampton. He was worried about the stock market; he tended to become panicky if he was left alone. Talking to an acquaintance on 13 June, when the ship was two days south of Madeira, he murmured something about blood poisoning but refused to elaborate on his remark. That same evening he wandered into the cabin of another passenger and offered him what appeared to be a small bag of diamonds. The man laughed and refused to accept the stones.

The following morning Barney played chess with Solly. At luncheon he appeared quite gay. He ordered champagne and afterwards insisted on Solly accompanying him on a walk round the deck. The sea was rough and after an hour or so of pacing up and down Solly grew tired and flopped into a deck chair. Barney continued his walking and then came to sit beside his nephew.

A small girl, Nellie Mackintosh, was one of the few eyewitnesses of what happened next. The incident made such an impression on her that she could recall it vividly many years later. She and her sister were friendly with the Barnato children, Leah and Jack, and were playing with them close to where Barney had sat down. 'He was sitting with another man in an alcove on the deck,' says Nellie Mackintosh in an unpublished statement. 'Suddenly he called out "They're after me", rushed to the nearby railings, climbed up and jumped over. As he went, the back of his coat blew up and over him; the suit was a dark brown with stripes in it and I can see it distinctly still. Then another man jumped over and I just saw him falling from above. The bells began to ring loudly and sailors ran in a long line, carrying blankets over one arm and a bucket in their other hand. A ladder went down and then a boat. Little Leah was clapping her hands. In the meantime people streamed on deck and all gathered at the place where the ladder was. I got so worried, thinking the boat might turn turtle, that I went to the other side and leaned hard over to try and help it to keep even! There was a heavy swell and we could just see the man bobbing away. I did not see him get into the boat.'

The man who leapt to Barney's rescue was the ship's fourth officer, W. T. Clifford. He had been dozing in a deck chair when, he was to say, he heard a cry: 'Murder! For God's sake, save him.' (Nellie Mackintosh is emphatic that there was no such cry.) Clifford also thought he saw Solly Joel making an attempt to clutch at Barney's trouser leg. What is more certain is that Clifford immediately dived overboard. He caught sight of Barney, who appeared to be swimming strongly, but the rough sea kept them apart. The *Scot* stopped. A lifeboat was lowered and it picked up Barney and Clifford. Barney was found floating face downwards: attempts to revive him by artificial respiration failed.

'We were told the next day', says Nellie Mackintosh, 'that Mr Barnato had died and we must be very quiet. I remember too when we reached England, the sailors carrying the coffin: it was roughly painted very black and had raw rope handles. Mrs Barnato, Leah and her little brother, both small, walked behind it. We had not seen them to play with from the day after his death as they had all stayed out of sight.'

At the coroner's inquiry, a verdict of 'Death by drowning while temporarily insane' was returned. As Clifford testified that Barney had been swimming strongly, the alternative verdict of deliberate suicide was ruled out. Clifford was commended for his bravery and was later rewarded by Woolf Joel and by public subscription. Lloyds presented him with a silver medal.

Barney was buried in the Jewish cemetery at Willesden on 19 June, 1897. He had died three weeks before his forty-fifth birthday. Some two hundred carriages followed his coffin. Alfred Beit, despite his dislike of Barney, was among the mourners.

It was a fitting end to a bizarre life. Barney can hardly be said to have chosen the way he died; his mind was too far gone for that. Yet, in some ways, his death was not inappropriate. A few months before he jumped from the *Scot*, Stuart Cumberland, a fashionable 'mind reader', published a book on South Africa. One of the chapters was devoted to Barney. Cumberland quotes several maxims passed on to him by 'the Great Barnato'. One of them could serve as Barney's epitaph. 'Always', he said, 'wind up with a good curtain, and bring it down before the public gets tired, or has time to find you out.'

[5]

When a rich man dies in unusual circumstances suspicions are always aroused. Inevitably there is talk of foul play. Barney Barnato's 'suicide' was therefore bound to spark off a spate of rumours: it would have been remarkable had it not. For, not only was he a rich man, he was a man whose life had been far from innocent. He had been involved in questionable transactions; he had made many enemies; he kept shady company. Practically everything he did was controversial: his death was no exception.

The theory that Barney Barnato was the victim of a deep-laid plot, rather than a man who had lost his reason, has often been advanced. Some members of his family were convinced of it. Fanny Barnato, for instance, is said to have clung to the belief that there was more to her husband's death than revealed by the coroner's inquest. While such a belief is understandable, there seems to be little evidence to support it.

That Barney's mind was deranged during the last months of his life there can be no doubt. He was schizophrenic; he had suffered repeated nervous breakdowns. Severe as was his neurosis, it was not uncharacteristic. Throughout his life he had been subject to brainstorms. In a crisis—particularly an emotional crisis—he was liable to react with a vehemence which startled even his close friends and associates. That these bouts of violent, irrational behaviour should intensify as he grew older is hardly surprising. He consistently refused the only advice that could have helped him. What he needed was a restful, placid existence, but he insisted on driving himself to extremes. He lived on his nerves, under constant pressure; he was seemingly incapable of relaxing, even when he could afford to do so. The pace he set for himself would have taxed the powers of the most integrated personality; for him—high-strung and basically unstable as he was—it proved disastrous. Doctors had warned him of what would happen. His friends had seen it coming. One of them had remarked that he was a bundle of quivering nerves and that 'some day that marvellous vitality will cease. Either life or brain will go.' The brain went: so did life. There can really be very little mystery about Barney Barnato's death.

What doubts there are arise largely from hearsay. They come from the evidence given by William Clifford at the coroner's inquest. Clifford said he was awakened by the cry of 'Murder!' Nellie

Mackintosh says there was no such cry. Admittedly Nellie Mackintosh was very young, but she was awake and Clifford was dozing. In any case, shouts at a time of crisis are often confused: they cannot be taken too literally. Then there was the fact that Barney was seen to be swimming strongly. But this proves nothing. It was admitted that Barney did not deliberately commit suicide but jumped overboard in a fit of temporary insanity: even an insane man can retain the instincts of self-preservation.

Exactly who is supposed to have shouted 'Murder!' is not clear. Some claim it was Solly Joel, others say it was Barney himself. Solly later remarked that immediately before Barney leapt from his chair he had asked the time. 'As I looked down to my watch,' said Solly, 'I saw a flash and he was over.' If there is any significance in his reference to a 'flash', it is difficult to know what it could be. There was no question of a shot being fired. There is no evidence of a knife being thrown. Barney's body was recovered and no wound was found on it. If someone was merely trying to frighten Barney, they could not have guaranteed that he would jump overboard. A flash on the deck of a ship could have been anything or nothing. If Solly Joel really suspected murder he would undoubtedly have done something about it. He would surely have demanded an investigation. For an investigation among the passengers and crew of a ship would swiftly have revealed anyone with a motive for killing Barney. But nothing was done and Solly's remark becomes meaningless.

But there were those who claimed that Solly had his own reasons for not demanding an investigation. It is said that Solly wanted to get rid of his neurotic uncle. According to this theory, it was Barney who shouted 'Murder!' and Solly who was responsible for his cry. But what could Solly possibly have done to make Barney jump overboard? And how could he be sure that his uncle would drown? Barney clearly ran to the rail, climbed it, and jumped: he was not pushed. Moreover, he started to swim and might have been rescued. The ingenuity and luck that would be required to carry out such a murder, in broad daylight, makes the suggestion quite impractical. Had Solly wanted to get rid of Barney there were safer and surer ways for him to have done so.

If the rumours surrounding Barney's 'suicide' had been left to the gossips, they would doubtless have died a death of their own. There was, on the face of it, nothing to keep them alive. Shocked as most of his friends were by the news, few of them seriously suspected murder. Barney's family might have derived some solace from the thought of a mysterious assassin on board the *Scot*, but his friends were more realistic. Cecil Rhodes was positively callous. He was crossing Bechuanaland by train when the news reached him. The telegram arrived late at night and his secretary waited until the following morning before handing it to him. 'I suppose you thought this would affect me and I should not sleep,' Rhodes growled. 'Why, do you imagine that I would be in the least affected if you were to fall under the wheels of this train now?' Rhodes liked to sound tough. It has been suggested that he was simply trying to disguise his real distress. This could be true; sentiment was not unknown to him. On the other hand, he was never overfond of Barnato. He and Alfred Beit did, however, send Fanny Barnato the full share of yearly profits due to Barney as a life governor of De Beers. The life governors' dividend was not declared until some days after Barney's death and Rhodes and Beit could, had they been mean-minded, have legally claimed Barnato's share. It amounted to £30,000.

It was not Barney's associates who kept alive the rumours that he had been murdered or driven to take his own life. They accepted his suicide and went about their business. What encouraged the doubts about Barney Barnato's death, was its strange sequel.

For precisely nine months after he leapt from the *Scot*, one of his nephews was shot dead. The circumstances in which Woolf Joel was killed on 14 March, 1898, were truly mysterious.

[6]

At the beginning of 1898, Solly Joel received a threatening letter. The anonymous writer demanded a loan of £12,000. If this money was not paid, he said, he would be willing to admit all the blame for removing Solly 'to a better world, this or the other side of the river Styx, where Barnato may be glad to see you again'. Signing himself 'Kismet', he told Solly to place an advertisement in the personal column of the Johannesburg *Star* indicating that he was prepared to pay the money demanded. Once the advertisement appeared, instructions would be sent for handing over the money.

Threatening letters were no novelty to members of the Joel family. Solly, like most other rich men, had received his full share of them. Usually they could be recognised and dismissed as the fantasies of cranks. The 'Kismet' letter, however, fell into a different category. It was long and literate and, despite a tendency to ramble, it gave the impression of having been written by a desperate man. Solly took it seriously.

Acting on advice, he consulted Bob Ferguson of the Johannesburg police. Ferguson suggested that an advertisement be put in the *Star*, inviting 'Kismet' to call and see Solly. This was done but the letter-writer refused to reveal himself. Instead he wrote a series of letters repeating his demands and hinting that he was in possession of a political secret which Solly could use to his own advantage. The mention of politics is said to have scared Solly. His recent imprisonment for involvement in the Jameson Raid fiasco had made him wary of political entanglements. He went to his brother, Woolf, for advice. Woolf then took the matter over. An advertisement was put in the *Star* telling 'Kismet' that any further arrangements would have to be made through Woolf Joel. Solly was packed off to join his family at the Cape.

Woolf Joel was undoubtedly the most intelligent, the most astute and the most respected of Barnato's nephews. His calm, good judgement had been invaluable to Barney in the negotiations preceding the amalgamation of the diamond mines. He had done much to further the Barnato interests in South Africa. In the years immediately prior to Barney's death, Woolf had spent a good deal of time in England. He had returned to South Africa in January 1898 as a managing partner in the Barnato firm. It is his reputation for reliability and sound sense that makes his handling of the 'Kismet' affair difficult to understand.

The threatening letters continued to arrive. They were, however, less demanding. The amount asked for dropped to £4,000 and Woolf was promised information which would benefit his firm financially. A clue to the writer's identity was also provided: he now signed himself 'Baron v. Veltheim'.

Woolf confided in Harold Strange, the manager of the Barnato concern. After discussing the letters, they decided that the police should be kept out of the affair. It is said that they were anxious to avoid any publicity that might harm the firm and possibly endanger Solly's life. But since they were dealing with a man who was obviously a potential criminal, any publicity resulting from his exposure could hardly have damaged the firm's reputation. Solly himself had already consulted the police and only the capture of his persecutor could really ensure his safety. The decision to act without the assistance of the police is puzzling. For Woolf Joel it proved fatal.

A meeting was arranged, at the request of the so-called Baron von Veltheim, in Barnato Park. Woolf and Harold Strange kept the appointment alone. They discovered von Veltheim to be an immaculately dressed man, over six feet tall, who spoke with a distinct German accent. The story he had to tell was astonishing. He claimed that, acting on behalf of a powerful group of insurrectionists, he had had several interviews with Barney Barnato before his death. A plan had been made to kidnap President Kruger and overthrow the Transvaal government. Not only had Barney been prepared to finance this plan but had offered up to £50,000 if it succeeded. With Barnato's death the plan had fallen through. However, he now had knowledge of a new plan and if the Joels were wise, they could profit from it. There was to be a *coup* in which Kruger would be toppled and replaced by a more accommodating President. All Woolf had to do was to sell his stock while the market was still good and then re-buy when prices fell after the overthrow of the government. With this advance information it would be possible for Woolf's firm to make millions. When the question of payment was raised, Woolf hedged. He asked for time to think the matter over.

Now, surely Woolf should have gone to the police. If he had nothing to hide, his course was clear. He had identified the letter-writer. It would have been a simple matter to arrange for von Veltheim to be trapped. Once the man was behind bars his brother would be safe. To unmask a cranky revolutionary (as von Veltheim must have appeared) could only reflect credit on his firm. Unless von Veltheim had made other threats during this interview—threats that have never been disclosed—Woolf's subsequent behaviour seems inexplicable.

For he did not go to the police. He continued to negotiate with von Veltheim. Exactly what there *was* to negotiate about is obscure. Von Veltheim had disclosed his great political secret. Woolf Joel was too intelligent a man to suppose he could bargain for Solly's life with such a preposterous adventurer. Even had there been any truth in von Veltheim's wild story—which is most unlikely—the Joels could hardly be blackmailed for Barney's behaviour, particularly in the last months of his life. However—whatever it was about— bargain with von Veltheim he did. Not that the Baron was such a difficult man to come to terms with. On the contrary, he was most conciliatory. Soon he had dropped his price to £2,500; hinting that he was prepared to abandon the revolution and depart in peace for England. When this was not forthcoming he wrote asking for £200 to help one of his fellow revolutionaries. This prompted Woolf Joel to declare that von Veltheim could be 'squared for a fiver'. Whatever he might have meant by 'squared', he was wrong.

On Sunday 13 March, 1898, von Veltheim contacted Woolf Joel and demanded an interview at ten-thirty the following morning. Reluctant to prolong the business, Woolf arranged for Harold Strange to meet the Baron that same afternoon. They met in the street. Strange told von Veltheim that Woolf was prepared to discount a promissory note for £200, but would offer nothing more. It was more than the fiver that Woolf had considered sufficient, but it was not enough.

The next morning, when Strange entered the Barnato office block, he was followed by von Veltheim who had been waiting outside. The Baron insisted on seeing Woolf Joel. Strange, armed with a revolver, took him to Woolf's office.

Woolf was sitting at his desk, his back to the wall. Von Veltheim came straight to the point. He demanded £2,500. Woolf repeated his offer to lend him £200. This infuriated von Veltheim. 'If that is your final decision,' he snapped, 'you know too much, and neither of you will leave this room alive.' Alarmed at his tone, Strange reached for his gun. The Baron was too quick for him. Pulling out a revolver and telling Strange not to move, he fired—first at Strange, then at Woolf Joel. Strange ducked and fired back. Woolf slumped across the desk, hit by three bullets. He had been reaching for his own gun when he was shot and had dropped it on the desk. Strange, seeing Woolf's gun, made a dive for it; he was stopped by the Baron stepping on his thumb. As von Veltheim fired again, the door burst open and some members of the staff rushed in and pulled him to the floor.

Woolf Joel was dead.

[7]

Von Veltheim was tried for murder. In the course of the trial much of his disreputable past came to light. He made many wild claims for himself and it is difficult to disentangle facts from his fantasies. Enough is known about him, however, to leave no doubt about his talent for imposture.

His name, of course, was not von Veltheim. Nor was he a Baron.
He had been born at Allhausen in Brunswick on 4 December 1857
and his name was Karl Frederic Moritz Kurtze. After deserting from
the German navy as a young man, he had served in various British
ships under a number of aliases. He first appears to have styled
himself Baron von Veltheim in Perth, Australia, where he married
for the first time. Leaving his wife and Australia, he wandered about
for some time in the United States and South America. Before
arriving in South Africa, he is known to have contracted at least two
bigamous marriages, both of which involved him in some extremely
shady transactions.

When Barney Barnato returned to England in 1896, weighed down
by the recent Reform Committee trial and his financial worries, von
Veltheim was staying at the Hotel Metropole in London. The
Jameson Raid was still very much in the news and von Veltheim
could well have contacted Barney with his melodramatic plan to
kidnap President Kruger. He was to assert that he did. Moreover,
he claimed that Barney agreed to the plan and promised to finance it.
A year later, on 17 April 1897, von Veltheim sailed for South Africa
on the s.s. *Ionic*. His passage was booked under the name of Franz
Louis Kurt. He arrived at the Cape about the same time as Barney,
who had just been sent from Johannesburg to recuperate before
returning to England on the *Scot*. Again the two of them could have
met; again von Veltheim said they did.

It is doubtful whether there was any truth in von Veltheim's
assertion that Barnato was financing him and had persuaded him to
come to the Cape. The story told by the self-styled Baron is full of
contradictions; it depends entirely on his own word and parts of it
are easily disproved. However, the occasions on which he claimed to
have met Barney are not without significance. They coincide with
important stages in Barney's mental breakdown. In the summer of
1896—when the first meetings are supposed to have taken place—
Barney became so obviously ill that his friends urged him to take
a holiday. At the Cape in 1897, he suffered a serious relapse and
shortly afterwards threw himself from the *Scot*. If von Veltheim's
account of his association with Barney Barnato is full of discrepan-
cies, it might well be because he was afraid to tell the truth. Barney
was desperately worried in the months before his death. Von
Veltheim could have been the cause of that worry.

A month or so after Barney's death, von Veltheim enrolled in the

Cape Mounted Police at Kimberley. He signed on for three years; he did not last three months. News of an unfortunate happening in London prompted the police authorities to ask for his resignation. In September, the naked body of a powerfully built man was washed up in the river Thames. One of the women von Veltheim had 'married' identified it as the body of her missing husband. Her mistake was soon discovered but the publicity given to the incident drew attention to von Veltheim. His photograph, and a report of his shady activities in London was sent to South Africa and the authorities decided that he was not a desirable police recruit.

He then drifted to Johannesburg. Here he stayed at a boarding house in Bok Street where he became friendly with a man named Caldecott who had lost a great deal of money by investing in some of the less successful ventures of the Barnato family. Mr Caldecott's daughter loathed Solly Joel; she blamed him for all her father's misfortunes. According to von Veltheim, this young woman was largely responsible for his writing the first 'Kismet' letter. How true this is one does not know. What is certain is that shortly after his arrival in Johannesburg he embarked on his campaign against the Joels.

There was undoubtedly more to von Veltheim's association with Solly and Woolf Joel than was ever revealed at von Veltheim's trial. Woolf Joel's handling of the affair makes this obvious. His behaviour was not that of a man threatened by a political crank, but of a man who is being blackmailed. Whatever political issues were involved would have incriminated von Veltheim far more than the Joels and hardly constituted a basis for blackmail. The precise nature of the hold that von Veltheim had over Woolf Joel remains a secret. In his threatening letters von Veltheim had been careful not to be too specific; he was equally guarded throughout his trial. It paid him to be so. By posing simply as a tool of unscrupulous capitalists who were intent on overthrowing the Transvaal government, he stood a better chance of gaining sympathy than he did by exposing himself as a blackmailer. He was playing a dangerous game, but it succeeded.

Despite his very dubious credentials, von Veltheim had his plea of self-defence upheld by the jury. 'I must say,' commented the judge, 'I am astonished at the verdict and do not agree with it.' Others in the court were not so surprised; the verdict was greeted with wild applause. The lawyer whom von Veltheim had employed to defend him, however, showed little enthusiasm. When his client came to thank him, he refused to take his hand. 'I got you off,' he said, 'but I don't shake hands with you.' Nor was President Kruger pleased to have his would-be kidnapper set free. Immediately after the trial the President ordered von Veltheim's re-arrest and deportation.

Did the bogus Baron have anything to do with Barney Barnato's death? Indirectly he could have done. There is ample evidence to show that Barney became hysterical when his nephews were threatened. It had happened when Isaac was arrested on an I.D.B. charge and again when Solly was imprisoned after the Jameson Raid. If von Veltheim had, as he seems to have had, some sort of hold over the Joels he might well have approached Barney first. Barney's devotion to his nephews was well known and he was thought— mistakenly—to be the richest member of his family. This would have made him an obvious target for a blackmailer. The events of von Veltheim's alleged association with Barney Barnato closely coincide with Barney's mental breakdown. It is not stretching credibility too far to imagine that there could have been some connection between the two occurrences. Coming on top of his other worries, blackmail might have been the last straw for Barney Barnato.

[8]

The Joel family had not heard the last of Baron von Veltheim. On 11 June, 1907, Solly Joel—then in London—received another letter from his sinister persecutor. The letter was postmarked Odessa, Russia, and dated 6 June. It demanded a financial settlement of the Baron's 'outstanding account'. Another letter arrived the following month, sent from St Petersburg, informing Solly that von Veltheim would shortly contact him by proxy. This proxy, a Mr Bumiller, arrived in London from Antwerp in September, bringing with him a draft order for £16,000. He refused to give any information concerning von Veltheim's whereabouts.

This time Solly acted in a more resolute manner. He could afford to. Not only was he now the powerful head of the Barnato concern— and popularly known as the 'Ace of Diamonds'—but he was dealing with a man who had admitted to killing his brother. Such a man, turning up again with threatening letters, could expect no sympathy from a jury. Nor was there much chance that he would make any

damaging disclosures. For if von Veltheim had refused to acknowledge the true relationship between himself and the Joels when he
was on trial for his life, it was doubtful whether he would do so on a
lesser charge. Solly did not hesitate to hand the matter over to the
police. Mr Bumiller was followed and von Veltheim was traced to
Paris where, on 19 September 1907, he was arrested.

Von Veltheim was brought to trial at the Old Bailey on 9 February
1908. As the threatening letter he had sent to Solly from Odessa
had referred vaguely to 'the history of the past' the prosecution was
allowed to recapitulate the events that had led to the shooting of
Woolf Joel. Once again the story of political plots and the kidnapping
of President Kruger, served as convenient red-herrings. The
evidence given at this second trial centred largely on von Veltheim's
claim to have been employed by Barney Barnato as a political agent.
Why this should have emboldened him to renew his attempt to
blackmail Solly Joel nine years later was never made clear. The only
man who, with the passing of years, might have shed some light on
the mystery was Harold Strange but the defence did not call him as
a witness. Von Veltheim's own evidence was as confused and as
misleading as ever. He tried to place the blame for his first black-
mailing attempt on the unfortunate Miss Caldecott who, he claimed,
had originally given him the idea. (This young woman had conveniently died during his South African trial.) The jury at the
Old Bailey were not impressed by his muddled explanations. After
less than twenty minutes deliberation they found him guilty of
attempted blackmail. He was sentenced to twenty years' imprisonment.

He did not serve his full term. Seven years later he was released in
reward for saving a warder's life during a prison break. In 1929 he
returned to South Africa in search of money. The money this time
was the treasure which President Kruger was supposed to have
buried before he went into exile during the Anglo-Boer war. Von
Veltheim claimed to know where the 'Kruger millions' were hidden.
However, his quest did not last long. While staying in a Transvaal
hotel, he was arrested and once again deported. He died, penniless,
in Hamburg the following year.

A further tragedy occurred in the Barnato family a few months
after von Veltheim's second trial. At the end of November 1908,
Barney's brother, Harry Barnato, died unexpectedly of a cerebral
paralysis. He was only fifty-eight.

As the flamboyant 'Signor Barnato', Harry had been largely responsible for his family's involvement in South Africa. First to arrive at the diamond diggings, he had for many years completely overshadowed his younger brother. Diggers looking back to those early days often confused Harry with Barney. They remembered a showy, much talked about conjuror named Barnato and imagined that it must have been Barney. In later years Barney had several times to point out that he had never been a stage juggler in Kimberley. The mistake, however, is understandable. From the time that Isaac Joel fled the country, Harry Barnato had spent less and less time in South Africa. He seemed reluctant to draw attention to himself. That he was by no means a colourless personality is evident from his career, but he appears to have been content to remain in the background and let Barney steal the limelight.

In the more spectacular Barnato ventures, Harry played an inconspicuous part. Although he had little or nothing to do with the amalgamation of the diamond mines, he was undoubtedly proud of his brother's achievement. One of his most cherished possessions was the famous cheque for £5,338,650 paid for the Kimberley Mine. He had the original cheque framed and kept it in his office. Occasionally he visited the Rand and, like the rest of his family, came in for his full share of blame for the financial disasters—such as the failure of the Barnato Bank—which dogged Barney's last days. But, for the most part, Harry Barnato preferred to keep out of the public eye.

He had a house at Nice, in France, and spent a great deal of his time there. His only enthusiasms, outside of business, were for horseracing and his daughter, Lily. He owned several horses and his greatest Turf triumph came when one of them, Sir Geoffrey, won the Lincoln Handicap. On his daughter he doted. She was his only child and her marriage to Samuel Asher in March 1903 moved Harry to one of his rare acts of generosity: to mark the occasion he donated £500 to various charities. When he died he left Lily a million, plus £10,000 a year.

Rightly or wrongly, Harry Barnato had a reputation for meanness. In March 1906 the racing scandal-sheet, *The Winning Post*, published a scurrilous article—probably inspired by Lou Cohen—attacking Harry. Besides raking up the old imputations of I.D.B., it condemned him for being tight-fisted. 'With the exception of a profuse exhibition of precious stones and an odd diamondiferous dinner, you do not live ostentatiously,' Harry was told, 'and had you a fourth, aye, a tenth of your wealth, the interest would overlap your expenses. Expenses? You spend little, and we can only discover that you give less. . . . There are many Jewish charities badly in need of money. There are heaps of poor miserable creatures who live within the precincts of your birthplace, who are sadly crying out for a helping hand. The opportunity of doing good is open to you at every corner.'

What effect this had on Harry one does not know. Unlike his nephew Jack (Isaac) Joel, he was wise enough not to sue the paper for digging into his past.

Rumours of Harry's wealth were wildly exaggerated. When he died the South African papers reported that he had left twelve millions. This proved to be untrue. But he was undoubtedly the richest of the Barnato family. His estate was later estimated at over £5,800,000. The publication of his will did much to dispel the accusations of miserliness. Besides the large sum bequeathed to his daughter, he left £250,000 to charity. His bequests were largely to medical institutions and included an endowment at the Middlesex Hospital to commemorate the tragic deaths of Barney Barnato and Woolf Joel.

Jack Joel proved to be the next richest member of the Barnato 'diamond dynasty'. When he died in 1940 he left an estate of £3,600,000. Woolf Joel was worth £1,226,600 when he was shot. The most extravagant of the three brothers, Solly Joel, was hit by a share slump shortly before he died of a heart attack in May 1931. All the same, his estate was valued at over a million.

Surprisingly enough, Barney Barnato did not die a millionaire—not quite. There was some foundation for his hallucinations of great financial losses. His desperate battle to bolster the share market, in the years preceding his death, had cost him dear. At the time of his death, his personal estate was valued at precisely £960,119 2s. 3d. He was a very rich man, but he should have been much richer. Had he not died when and as he did, there can be no doubt that he would have been.

Barney's greatest talent was for making money. But his career cannot be assessed entirely by the amount of money he made. He started life with many disadvantages and fought his way to the top. Had he died penniless his achievements would have been remarkable. The competition was tough, his opponents ruthless. If he was also tough and ruthless it was because he had to be. 'Never', he once said, 'let a man put his hand on you without giving him "what for", and always have the first hit.' His tragedy was that the last blow was struck when his defences were down.

CHAPTER TWENTY

THE FALL AND DECLINE OF RHODES

SHORTLY before J. B. Robinson left for the Transvaal in July 1886, a merchant named Fred Alexander returned to Kimberley with specimens of gold-bearing rock. His arrival created much excitement. He had picked up the rock on the Witwatersrand and, to prove that it was auriferous, he invited several prominent Kimberley men to see a panning. The demonstration took place behind Alexander's shop on 16 July. It was a great success. News of the telegram telling Robinson of the discovery of gold had already leaked out and the evidence provided by Alexander's samples helped to confirm the rumours. 'It may be relied upon that the nature of the gold fields is of a particularly attractive nature,' reported the *Diamond Fields Advertiser*, the day after the panning, 'or else Mr J. B. Robinson would not be going off to the new Eldorado. He leaves on Sunday for the Transvaal.'

Even so, these encouraging signs did not impress everyone. There had been too many unsuccessful gold rushes to start a new stampede from Kimberley. More than the panning of a few rock samples was needed to overcome the scepticism with which many of the old hands in Kimberley greeted the news from the Witwatersrand. Not least among them was Cecil Rhodes.

Rhodes is said to have been one of those who witnessed Fred Alexander's demonstration. If he was, it did not inspire him to follow J. B. Robinson. He had neither the time nor the inclination to go looking for gold. In July 1886 he was still busy trying to amalgamate the diamond companies of the De Beers Mine and was already toying with plans for the unification of the four Kimberley mines. Only a few months earlier he had battled to prevent J. X. Merriman from launching a rival amalgamation scheme; the competition was still too strong for him to be distracted by rumours of gold. Moreover, he had a personal reason for not wanting to leave Kimberley. His closest friend, Neville Pickering, was seriously ill.

Although Pickering had played little part in Rhodes's financial and political undertakings, he undoubtedly had a great influence on Rhodes's life. In 1886 they were still sharing the same corrugated iron cottage and Pickering was still Rhodes's sole heir. Their friendship was as firm as ever. Indeed, had it not been for his ill-health Pickering might well have played a more active role in Rhodes's schemes. As secretary of the De Beers Mining Company he had proved extremely efficient; he was well thought of among Rhodes's business associates. F. Philipson-Stow described him as 'certainly one of the best brokers for the Company's diamonds at Kimberley'. Unfortunately he was given little time in which to prove his usefulness. After living with Rhodes for two years he met with an accident which crippled him for life.

While riding in the veld, in June 1884, Neville Pickering was thrown from his horse. He fell into a clump of thorn bushes and was badly cut and bruised; 'some of the thorns,' it was reported, 'entering below the knee of both legs.' The poison from the thorns, added to the after-effects of a recent bout of 'Kimberley fever', weakened him considerably. For over a month Pickering was confined to bed; when he did get up he was forced to hobble about on crutches. He never fully recovered; his health was permanently impaired.

Throughout his friend's long and depressing bouts of illness, Rhodes's devotion was unquestioned. All that could be done to ease Pickering's suffering, he did. He nursed him and obtained the best medical advice. One of the doctors called to attend young Pickering was the popular and successful Scotsman—Dr Leander Starr Jameson. Although Rhodes appears to have known Jameson reasonably well for some years, it was not until Pickering's illness that they became good friends. It was a friendship which was to have important consequences.

When the news of the Witwatersrand gold finds reached Kimberley, Neville Pickering had just returned from staying with his family in Port Elizabeth. He was far from well. Had any persuasion been necessary to convince Rhodes that it was not worth forsaking his diamond interests to chase after gold, his friend's health would undoubtedly have provided it. For the first two weeks of the rush to the Transvaal, Rhodes remained in Kimberley.

The local newspapers continued to publish glowing accounts of the Witwatersrand. Reports of diggers flocking to the gold fields from all over South Africa, of the opening up of new farm areas and of the scramble for claims, appeared every day. There could be no doubt about the extent and importance of the diggings and it was obvious that those on the spot were losing no time in snapping up promising properties. J. B. Robinson had been followed by a few other Kimberley capitalists and their activities were reported in detail. Within a matter of days several new syndicates and companies had been formed. The situation was similar to that at New Rush some fifteen years earlier, when Herbert Rhodes had secured a foothold in the diamond diggings. That there was now nobody to represent the family interests in the Transvaal seemed not to bother Cecil Rhodes. It was not until the beginning of August that his attitude changed.

On 30 July 1886, the *Diamond Fields Advertiser* carried a long report on the doings of various Kimberley men on the Witwatersrand. Much of it was concerned with the valuable purchases made by J. B. Robinson. At the end of the report it stated that Dr Hans Sauer had just left for Kimberley 'with lots of specimens of auriferous soil picked up casually from Robinson's ground'. Rhodes might not have noticed this item, but he was soon to be aware of Dr Sauer's return.

Hans Sauer had become well-known in Kimberley for his fight during the smallpox epidemic. This had not endeared him to the diamond magnates. His concern with the gold fever, however, was to place him firmly in the capitalist camp. Sauer had been one of the first Kimberley men to leave for the Transvaal; he had, in fact, accompanied J. B. Robinson to the Witwatersrand. After trailing round the gold fields with Robinson he had become convinced of the area's potential but, not having sufficient capital, he had been unable to join in the scramble for claims. He returned to Kimberley hoping to obtain financial backing. His brother-in-law suggested that he approach Cecil Rhodes.

At that time Sauer hardly knew Rhodes. He was given a rather cool reception on his first visit to Rhodes's cottage. It was early in the morning and Rhodes was still in bed; Sauer sat on the edge of the bed and told his story. 'He listened to what I had to say about my journey and the gold deposits on the Witwatersrand without much apparent interest,' says Sauer, 'and when I had finished my tale he simply said: "Please come back here at one o'clock and bring your bag of samples." ' Returning to the cottage, Sauer was surprised

to find Charles Rudd and two Australian miners awaiting him. The
miners, who had brought gold-panning equipment with them,
quickly confirmed that the doctor's samples were indeed gold-
bearing. Neither Rhodes nor Rudd seemed particularly impressed.
However, Sauer was told to call on Rhodes that afternoon at the De
Beers offices.

During this afternoon interview Rhodes displayed slightly more
interest. He agreed to allow Sauer to act as his agent on the Witwatersrand and to give him a fifteen per cent interest in any claims
he acquired. The doctor asked for £200 to cover his immediate
expenses and it was arranged that he should leave for the Transvaal
the following morning. He was authorised to draw on Rhodes for
any reasonable amounts he might require for the venture.

The following morning Rhodes stopped Sauer in the street and
advised him to board the Transvaal coach at a stage outside of the
town. 'You see, everyone in Kimberley knows you only returned
from the Transvaal the day before yesterday, and they will be
wondering why you are going back there . . .', said Rhodes. 'It is
better not to excite curiosity.' Sauer did as he was told. He drove in
a buggy to the second coaching stage—some twenty miles from
Kimberley—and waited for the Transvaal coach. When it arrived
he climbed in and glanced round at his fellow passengers. To his
amazement he found Rhodes and Rudd seated in the two best
corner seats. 'No explanation was offered by either of them,' he says.

The story of Rhodes's and Rudd's sudden change of mind became
one of the jokes of the Witwatersrand. It was said that they had paid
a £3 premium for their coach seats, after having scoffed at Dr
Sauer's samples. 'Needless to remark,' says one report, 'they were
subjected to a surfeit of sportive raillery.' Before long they were
supplying the diggers with a good deal of amusement.

Rhodes seemed completely out of his depth on the Witwatersrand.
He had made himself a diamond expert but of gold he knew nothing.
Nor was Rudd much help to him. If anything Rudd was more wary
of buying the gold properties they were offered than was Rhodes.
Dr Sauer, who rushed about securing options for the partners, was
driven nearly frantic by their hesitancy. They bought some valuable
claims but others they refused to consider.

Even the arrival of Alfred Beit—who came to see how Robinson
was handling their joint syndicate—did nothing to inspire confidence
in Rhodes and Rudd. A particularly promising block of claims which

Sauer was offered for £500 was snapped up by Beit for £750. On hearing that Rhodes had recently turned the claims down, Beit was distressed. He told Sauer that he was prepared to give Rhodes a half interest in the block at cost price. But still Rhodes was not interested. The claims were then incorporated in the Beit syndicate and later formed part of the Robinson Gold Mining Company—one of the most valuable mines on the Witwatersrand. This was only one of many similar *gaffes*.

'Rhodes,' says Sauer, 'like most men in South Africa then, knew nothing of gold mining, and still less of gold-bearing ore bodies, and, in the back of his mind was the fear that the whole thing might turn out to be a frost. If he had taken up all or the greater part of the properties which I had secured under option for the matter of a few thousand pounds, he would undoubtedly have become one of the richest men that has ever lived.'

Whether things might have been different had Rhodes stayed longer on the Rand, one does not know. As it was, he gave himself little time to study the reef. He arrived in the Transvaal at the beginning of August and left before the end of the month. A message arrived to say that Neville Pickering had taken a turn for the worse: he was thought to be dying. To Sauer's astonishment Rhodes announced that he was returning to Kimberley immediately. When Sauer protested that his signature was needed to close some important deals, Rhodes refused to listen. 'I'm off,' he declared.

The coach that evening was fully booked. This did not deter Rhodes. 'Buy a seat from someone who has already booked,' he said. 'Get a special coach—anything; I am going tonight.' None of the passengers was prepared to give up a seat but in the end a place was found for him on top of the coach; he completed the three hundred dust-choked miles propped up among the mail bags. The journey lasted for more than fifteen hours.

His haste was not strictly necessary. Neville Pickering lingered on for a few more weeks. Rhodes never left his side. It was said that he was 'careless of anything but the wants and comforts of his friend'. At times he was joined by Neville's brother William, who was then the manager of a bank at Dutoitspan. Together they sat and watched the life ebb from this once virile, active and light-hearted young man.

A legend was to grow around the vigil that Rhodes kept at his friend's bedside. It was said that he refused to attend to any business;

that frantic telegrams from the Rand went unanswered; that by
neglecting Sauer's options he destroyed his chances of controlling
the gold industry. This is only partly true. He did keep in touch
with his Transvaal agents and he had earlier turned down several of
Sauer's options. All the same his unquestioned devotion cannot be
lightly dismissed. There can be little doubt that his concern for
Pickering was greater than his interest in the Witwatersrand. Rhodes
might not have sacrificed a fortune by returning to his friend, but he
was obviously prepared to do so.

The end came on the morning of 16 October, 1886. Dr Jameson
had been sent for a few hours earlier but there was nothing he could
do. Shortly before he died, Neville managed to stir himself. Turning
to Rhodes, he whispered: 'You have been father, mother, brother
and sister to me.' He died in Rhodes's arms.

All Kimberley turned out for the funeral. Pickering had been well
loved and the sorrow caused by his death was sincere. The chief
mourner was his brother William, but the most conspicuous was
undoubtedly Cecil Rhodes. Dressed in a crumpled old suit, passing
from tears to hysterical laughter and burying his face in a large
handkerchief, Rhodes hardly seemed to know where he was or what
he was doing. As he turned from the grave, he came face to face with
the sobbing Barney Barnato. 'Ah, Barney,' he cried, 'he will never
sell you another parcel of diamonds.'

Such unhibited displays of emotion were rare for Rhodes. After
Pickering's death they became rarer. He prided himself on his
cynicism, his toughness and his realism. He took accusations of
ruthlessness as compliments. There was no place in his scheme of
things for tenderness. He rarely, if ever, spoke of his dead friend.
But his grief was as genuine as his silence was significant.

[2]

After Neville Pickering's death, Rhodes moved from the cottage
they had shared. He went to live with Dr Jameson in another little
corrugated iron house opposite the Kimberley Club. His way of life
did not change much. Jameson was a successful doctor and Rhodes
was a diamond magnate, but they lived like impoverished diggers.
The house was sparsely furnished and always untidy; there were
two bedrooms containing little more than truckle-beds, and a
sitting-room which looked, according to one astonished visitor,

'like that of an undergraduate at college'. Most of their meals were eaten at the Club across the road.

It was in Dr Jameson's house that Rhodes held that all-night session—with Beit, Woolf Joel and Barney Barnato—to determine the terms of the De Beers Consolidated Mines trust deed. By then he was not only in control of the diamond mines but had gained a firm foothold on the Rand. For, after Pickering's death, Rhodes had returned to the Transvaal and made some profitable investments. At the beginning of 1887, he and Beit had visited the Witwatersrand and had tried to talk J. B. Robinson into an amalgamation of the gold mines. When Robinson—who held the whip hand—refused to cooperate, Rhodes had been obliged to make the best of what he already held. He incorporated his various gold mining interests into The Gold Fields of South Africa Limited—the company which he and Rudd founded in February 1887, with a capital of £125,000. Although this company by no means equalled the concerns controlled by Beit and Robinson, it more than compensated Rhodes for his initial tardiness. By 1892, when it was renamed The Consolidated Gold Fields of South Africa, its capital had been increased by a million and a quarter and three years later the dividend had risen to no less than fifty per cent. Rhodes made an enormous profit from the Gold Fields Company: for several years he drew some £300,000 to £400,000 from it annually. He also made sure that the Trust Deed of the company—like that of De Beers—allowed him latitude for his political designs.

These designs had been considerably furthered by the amalgamation of the diamond mines. Once he was financially assured and free of his battle with Barney Barnato, Rhodes had again been able to 'think in Continents'. More specifically he had been able to concentrate on his plans for occupying central Africa. His negotiations with Lobengula, the Matabele chief, had by that time reached a satisfactory conclusion.

On 30 October 1888—a few months after Barney Barnato had capitulated and agreed to the amalgamation of the diamond mines —Lobengula agreed to give Rhodes's agents 'exclusive charge over all metals and minerals' in his kingdom. This was the famous concession obtained for Rhodes by Charles Rudd. In return, the Matabele chief was to be paid £100 on the first day of each lunar month and supplied with a thousand Martini-Henry rifles and a hundred thousand cartridges. The promise of an armed steamboat on the Zambesi was thrown in for good measure. When Lobengula affixed his great elephant seal to this agreement it is doubtful whether he fully understood its implications. He is said to have expressed the wish that not more than ten men would come and dig for minerals in his territory as a result of his granting the concession. Rhodes, of course, had very different ideas.

It was shortly before the signing of the Rudd concession that the disgruntled shareholders of the Central Company had challenged the trust deed of De Beers Consolidated Mines in court. During the hearing of the case their lawyer had compared the new company with the old East India Company; all that De Beers lacked was a Royal Charter. When Rhodes came to form the British South Africa Company, which was to occupy Lobengula's territory, he determined to rectify this omission. In March 1889, he went to London to obtain a Royal Charter. His mission was successful. Soon after his return to Cape Town, later that year, it was announced that Her Majesty Queen Victoria had been pleased to grant a Royal Charter of Incorporation to the British South Africa Company. Rhodes was now all set to add a considerable slice of Africa to the British Empire.

In June 1890, the Pioneer Column of the new Chartered Company marched into Lobengula's territory. The members of the column were mostly young men recruited in South Africa. Included among them were artisans and tradesmen who, upon arrival, would provide the various skills necessary for establishing a civilian settlement. Their journey north took over two months. They hacked their way through dense bush, waded across swamps, bridged rivers and marked new roads on their maps.

They were headed, not for Lobengula's own domains in Matabeleland, but for his vassal state of Mashonaland. Nevertheless, the Matebele chief regarded the invasion with apprehension; it was far larger than he had been led to expect and his young warriors had grown restive. Aware of this hostility, the Pioneer Column kept searchlights blazing at night to frighten off any contemplated attack. Towards the end of August the Pioneers emerged on the open plains of Mashonaland and on 11 September they came in sight of their destination. Tents were pitched, guns were fired and the flag hoisted. They named the place Fort Salisbury after the British Prime Minister and an express letter was sent to Rhodes.

'When at last I found that they were through to Fort Salisbury,' said Rhodes, 'I do not think there was a happier man in the country than myself.'

He had been unable to accompany the column. Much as he had wanted to, he had been held back by his political commitments. Shortly before the column set out, he had been summoned to Cape Town. A crisis had arisen in the Cape Assembly and he was determined to take advantage of it. Railway construction was involved; as he intended to build a railway line from the Cape to Cairo, it was a matter with which he was deeply concerned. He hurried from Kimberley to Cape Town to vote against the Government. This was the first time he had taken his seat in the Assembly for months but such was his reputation in the Colony that, when the Government duly fell, it was he who became the new Prime Minister of the Cape.

His parliamentary duties now ruled out any possibility of his administering the territory occupied by the Pioneer Column. This job eventually fell to Dr Jameson. It was not an easy one. The Matabele were hostile and, rightly, suspicious; and the white pioneers were soon casting predatory eyes on Lobengula's unoccupied domains. A clash between the Matabele and the white settlers became inevitable. It came (or was organised by the Chartered Company) in October 1893. Dr Jameson marched against Lobengula, occupied the chief's great kraal at Bulawayo and added Matabeleland to the Chartered Company's possessions. For Lobengula, defeat was quickly followed by death. He died in the bush—some say of a broken heart—soon after fleeing from his kraal. Jameson acted with the full concurrence of Rhodes; his war against the Matabele reflects little credit on either of them.

Rhodes now had both Mashonaland and Matabeleland: two territories which were to bear his name. For a long time the Chartered Company's lands had been known as Rhodesia, but it was not until the beginning of 1895 that this name was officially recognised. When Rhodes heard that the name had been approved, he was jubilant. 'Has anyone else had a country called after their name?' he exclaimed. 'Now I don't care a damn what they do with me!' He spoke too soon. The newly named country was partly responsible for the greatest crisis of his career.

[3]

On his brief visits to Rhodesia (he went whenever he could spare the time) Rhodes had been aggressively optimistic. He refused to listen to criticism of the country. But no amount of brave talk could disguise the fact that—whatever its potential—Rhodesia was not another Transvaal. It had not lived up to its early reputation as the new Eldorado.

Originally, great hopes had been held of uncovering untold mineral wealth in central Africa: particularly gold. Mashonaland had long been a land of legends. It was popularly known as the Land of Ophir; it was rumoured to be the source of the Queen of Sheba's riches, the location of King Solomon's mines. Reality had shown otherwise. No great mineral deposits were found in either Mashonaland or Matabeleland. Gold there was, but not in large quantities. The reports of mining experts were far from encouraging. Disappointment made Rhodes look elsewhere. It made him look towards the Transvaal. It helped propel him into the Jameson Raid fiasco.

There were, of course, other reasons for his wanting to topple President Kruger. From the time he had assumed the premiership of the Cape, Rhodes had worked for the unification of the states of southern Africa. This had always been an essential part of his great vision. Unless it were achieved his empire would be built on uncertain foundations; all his other schemes in Africa would be jeopardised. The greatest obstacle to such a union was, as he soon discovered, Paul Kruger. At every turn Rhodes found his plans frustrated by the shrewd, implacably hostile old President. Progress might be made in establishing ties with Natal and the Orange Free State, but there was little hope of making headway with the Transvaal while Kruger presided in Pretoria. Rhodes had long recognised the need to remove President Kruger. The question was how and when?

The poor mining prospects in Rhodesia helped to supply the answer. In September 1894, Rhodes engaged an American mining expert, John Hays Hammond—who had been brought to South Africa by Barney Barnato—to survey the Chartered Company's territories. Hammond's report was far from favourable. 'If he cannot say anything stronger than that,' remarked one of Rhodes's associates on reading the report, 'I have not much hope for the future of the Chartered Company.' The report could not have come as a surprise to Rhodes. He and Jameson had accompanied Hammond on his tour of the northern territories and had a good idea of what his findings would be. Mining was not the only problem discussed on this tour. Hammond was well acquainted with the grievances of

the uitlanders in the Transvaal and he spoke of those grievances
to Rhodes and Jameson. The combination of circumstances undoubtedly contributed to Rhodes's decision to support the uitlanders
in their quarrel with Kruger. A successful *coup* in the Transvaal
would have solved a great many of his problems.

For their can be little doubt that, had the Jameson Raid proved
successful, Rhodes intended to replace Kruger's grip on the gold
mining industry with a grip of his own. His grip would have been
more enlightened and more progressive, but it would have been a
grip nonetheless. Rhodes made no secret of this. His aim was not
merely to oust Kruger, but to control the gold mines. The idea of
replacing the Boer republic with an uitlander republic was anathema
to him. 'I hate Kruger,' he said, 'but I prefer him to J. B. Robinson
or Barney Barnato.' He wanted no further financial fights. What
he had manoeuvred in Kimberley, he was planning to grab in
Johannesburg.

The failure of the Jameson Raid stunned him. When he heard
that Jameson had been captured, he could hardly believe it. 'Old
Jameson,' he told a visitor, 'has upset my apple-cart. . . . Poor old
Jameson. Twenty years we have been friends, and now he goes in to
ruin me.' Rhodes had been among those who had tried to stop the
impetuous doctor at the last moment.

But Rhodes did not brood for long. His first thought, after
resigning as Prime Minister of the Cape, was for the safety of De
Beers. He went straight to Kimberley, where he was met at the
station by a cheering crowd. This loyal demonstration touched him.
'In times of political adversity', he told the crowd, 'people came to
know who their friends were.'

Once he had settled his affairs at De Beers, he returned to the
Cape and sailed for England. There were urgent matters he had to
attend to in London. Not only was his own involvement in the Raid
being spoken of, but it was rumoured that the British government
was implicated. Shortly before the Raid, Rhodes's agents had been
in close contact with the British Colonial Office. Several telegrams
had been sent between London and Cape Town which, it was said,
would prove that the British Colonial Secretary had connived with
Rhodes to overthrow the Boer government. Rhodes was determined
that the contents of these telegrams should remain a secret.

And a secret they did remain. At the subsequent inquiry into the
Raid—held in London at the beginning of 1897—these 'missing

telegrams' did much to obscure more vital issues. Rhodes freely admitted his responsibility for Jameson's incursion, but refused to disclose details of the plot which led up to it. So many conflicting interests were involved that the inquiry became as great a fiasco as the Raid itself. One of Rhodes's many critics summed it up wittily as 'The Lying in State at Westminster'. But the confusion worked very much to Rhodes's advantage. He was let off with little more than a censure.

His real punishment came later. For he never fully recovered from the disgrace of the Jameson Raid. It killed his hopes; destroyed his vision. He continued to control De Beers; he retained his interests on the Witwatersrand; he contributed greatly to the development of Rhodesia; but his political influence was sadly diminished. The best days were over.

Shortly after the Raid, a friend had bet that Rhodes would have regained his former prominence within a year. Told of this bet, Rhodes thought for a minute and said:

'It will take ten years; better cancel your bet.'

His prophecy was never put to the test. He did not live another ten years.

[4]

Unlike the other mining magnates, Cecil Rhodes never aspired to a Park Lane mansion. He, the most patriotic of Englishmen, did not even live in England. When in London, he invariably stayed at the Burlington Hotel in Cork Street, where the same suite of rooms was always available to him. He had his own private dining-room, a sitting-room and as many bedrooms as he required. Here he entertained his friends, conducted his business and endeared himself to the staff with his lavish tips. After one short stay, for instance, he handed his favourite waiter, Arthur Sawyer, a cheque for £75, saying, 'I am going away today—here is a little present for you.' He was always welcome at the Burlington. He thought of it as his London home.

The only real home he owned was at the Cape. In the suburb of Rondebosch, outside Cape Town, he bought a rambling old two-storeyed house, known as 'The Grange', built on the lower slopes of Devil's Peak. It was a place of historic interest and dated back to the days of the Dutch East India Company, when it had been used as a granary. Originally a delightful, whitewashed barn with muscular gables and a high-pitched thatched roof, it had lost much of its pristine charm by the time Rhodes became interested in it. A succession of owners and a series of shoddy conversions had effectively camouflaged the dignified simplicity of the original buildings.

Rhodes first took a lease on the house in 1891, shortly after he became Prime Minister of the Cape. After renting it for two years, he persuaded the owner to sell and commissioned Herbert Baker—then an unknown young architect—to restore its original architectural character. Working from a water-colour sketch of the old granary, Baker faithfully restored the house, which—given back its old name, Groote Schuur—became one of the show places of the Cape. It was one of the few personal possessions in which Rhodes took a real delight.

The house grew on him. At first he took only a perfunctory interest in the place. The furnishing he left to one of his secretaries, who solved the problem of filling the rooms by spending a quarter of an hour at Maples in London, ordering everything—from salt cellars to wardrobes—from a selection of 'three of each kind'. Happily this tasteless hotch-potch was soon thrown out by Rhodes. Once he had bought and restored the house, his interest in it deepened. 'The problem of house building and furnishing,' says Herbert Baker, 'so new to one whose life had been spent under the rough conditions of farm, veld and mine, then became of absorbing interest to him.'

He filled the rooms with old Dutch furniture, Gobelin tapestries, Delft china, copper, brass and African curiosities. Every detail was designed to reflect the lives of the early Dutch settlers. So keen was his sense of period that he even toyed with the idea of having no electric light, or even oil lamps, and burning nothing but tallow candles to keep the illusion of age. He had a great admiration for the robust qualities of the founders of the Cape. Their taste was similar to his own: chaste and masculine. 'The big and simple, barbaric if you like,' he instructed Baker. 'I like teak and whitewash.'

Rhodes's house was a man's house; only men lived in it. Sometimes he would have women guests and, on occasions, his sister Edith would visit the Cape and stay with him for a few months. But he did not like to have women about him for long. There were no women servants at Groote Schuur; even the wives of his male employees were kept at a distance. The maids of his guests were instructed to make themselves as inconspicuous as possible. His closest friends, the friends who stayed with him for long periods—men like Alfred Beit, Dr Jameson and Sir Charles Metcalfe—were all bachelors.

Rhodes's pointed avoidance of the opposite sex earned him the reputation of being a misogynist. His admirers, however, strongly deny this. 'This is a libel,' declared one during Rhodes's lifetime. 'True, he does not court women's society, and he is never found pouring empty nothings into a pretty woman's ear, but there are some women—women of the intellectual order—for whom he has an admiration. His aversion to women's society lies in the fact that he has little or no time to devote to them. . . . He agrees with Ibsen that the strongest man stands alone.' Rhodes liked to give this impression. When asked why he had not married, he would say he was too busy; that a married man should give care and attention to his wife and that he was far too occupied for such things. Few, if any, questioned this trite evasion. The Victorians were not concerned with the causes of sexual sublimation; a man such as Rhodes was regarded as being above the unmentionable temptations of the flesh.

But, of course, it was an evasion. Marriage is not something in which men indulge if they have time to spare. For most men it is the answer to a basic need. Other men have been as busy as Rhodes and still found time to satisfy their natural instincts; to mate and procreate. It has not been left to dedicated bachelors to shape the destinies of the world.

The reason for Rhodes not marrying went deeper than his preoccupations. Not only did he not take a wife but it is highly unlikely that he ever took a woman. Women did not attract him sexually. No woman ever came closer to him than his mother. Yet he was far from being a bloodless ascetic. He was a man of large—some say gross—appetites; he could, and did, display strong emotions. If his attitude towards women was austere, his emotions were very much in evidence in his relationships with young men.

'He seemed to have a liking for young men,' says one of his secretaries. As he became older this liking became more and more apparent. He would go out of his way to befriend any youngster who caught his eye; many of them were taken into his employ. A specially selected band—known as 'Rhodes's Angels' or 'Rhodes's Lambs'—were included in the Mashonaland Pioneer Column. Others were given jobs on his various estates in South Africa or Rhodesia. His particular favourites became his 'private secretaries'. More often than not these so-called secretaries were young men singularly unfitted to perform the simplest clerical duties. None of them could type, some could hardly write; when one of them took the trouble to learn shorthand, the rest regarded him as a freak. Their obvious incompetence drove Rhodes's business and political associates to despair.

'We were all much more companions than secretaries in the ordinary sense of the word,' admitted one of them. But they were hardly companions in the ordinary sense of the word. Rhodes had little in common with his 'secretaries'. He did not share his thoughts, his schemes, his hopes or his politics with them. What existed between them and Rhodes was an almost adolescent relationship: banter, horseplay and practical jokes. These hearty, often uncouth, young men could do what they liked with Rhodes; they teased him, bullied him and openly made fun of him. He tolerated their impertinence, indulged their whims, spoilt them, amused himself with them: but he never took them into his confidence. They provided him with a means of relaxation rather than companionship.

All that Rhodes demanded of his young male entourage was unswerving loyalty. They had to remain faithful; they had to remain single. The mere suggestion that one of them was contemplating marriage was sufficient to reduce Rhodes to hysterics. The scene he created when one of his secretaries became engaged was never forgotten by those who witnessed it. 'Rhodes raved and stormed like a maniac,' it is said. 'His falsetto voice rose to a screech as he kept screaming: "Leave my house! Leave my house!" No small schoolboy, or even schoolgirl, could have behaved more childishly than he did.' On the other hand, he was quick to return any devotion shown to him. He showered his favourites with gifts, nursed them through illnesses, and would not have a word said against them. So long as a young man remained true to Rhodes, he could do no wrong.

The intensity of his emotional involvement with these young men was such that it is difficult to escape the conclusion that Rhodes was homosexual. The attachments he formed, however, were purely platonic. Those who played the most prominent part in his life were ordinary heterosexual youngsters. It is extremely unlikely that any of them were conscious of anything untoward in Rhodes's feelings for them. Indeed, he might not have recognised the significance of his feelings himself. Overt homosexuality would have horrified him. His abhorrence of effeminacy was so excessive as to be suspect. It is said that he was so suspicious of jewellery on a man that he refused even to wear a watch. 'He liked a man to display the attributes of a man,' says one of his secretaries, 'and despised indecision, weakness, and effeminateness in the male sex.' The qualities he looked for in a

young man, would have precluded a reciprocal homosexual relationship. It is not unknown for repressed homosexuals to divert their
sexual inclinations into purely emotional channels.

To be surrounded by young men, to be the focus of their attention,
and to exercise control over their lives was, it would seem, sufficient
for Rhodes. But this was a definite emotional force in his life: women
could never arouse similar emotions in him. All his biographers
agree on one point: Rhodes did not hate women but he was decidedly
happier in the company of men.

It is therefore ironic that it was a woman who was held responsible
for his early death. This, at any rate, is what many of his friends
claimed. Lord Castlerosse, who learned about Rhodes from Dr
Jameson, was to write: 'Cecil Rhodes, though he never cared for
women or allowed himself to be influenced by them in the slightest,
was nevertheless killed by a woman.'

[5]

The woman was Princess Catherine Maria Radziwill. As far as is
known, she entered Rhodes's life about the same time that Baron
von Veltheim first claimed acquaintance with Barney Barnato—in
1896, shortly after the Jameson Raid. While there is doubt about von
Veltheim's association with Barnato, there can be no doubt about
Catherine Radziwill's pursuit of Rhodes.

Worldly, dark-haired and vivacious, the Princess was well-known
in European society. She was the only daughter of Count Adam
Rzewuski, an exiled Polish nobleman living in Russia. Christened
Ekaterina, she changed her name to Catherine Maria when, in 1873,
at the age of fifteen, she had married Prince Wilhelm Radziwill. The
early years of her married life had been spent with her husband's
family living in Berlin. Here she had gained a reputation both as an
outstanding beauty and as a sharp-tongued political meddler. Her
political intrigues at the German Court had eventually resulted in
her being banished from Berlin and, in 1886, she and her husband
had moved with their family—they had five children—to Russia. In
St Petersburg she had established a political *salon*, but her influence
in the Russian capital was diminished after the death of Tsar
Alexander III in 1894. The following year the Princess separated
from her husband. When she first met Rhodes she was eking out an
existence as a freelance journalist.

They met at a dinner-party given by Moberly Bell, the manager of the London *Times*, in February 1896. The Jameson Raid had taken place the month before and Rhodes had just arrived in England. The Princess appears to have been at her conversational best during the meal but Rhodes, then facing the greatest crisis of his career, was too preoccupied to notice her. When she later wrote to him, he had—or claimed he had—difficulty in remembering who she was.

They did not see each other again for three years. The Princess wrote to Rhodes a couple of times—once to send him a lucky charm and once to ask his advice about an investment—but otherwise they had no contact.

Their second meeting was on board the s.s. *Scot*—the ship from which Barney Barnato had jumped to his death—which sailed to South Africa from Southampton in July 1899. The Princess is said to have changed her booking in order to travel in the same ship as Rhodes. One of Rhodes's secretaries has given a colourful account of Catherine Radziwill's first entry into the ship's dining saloon. 'She glided into the saloon,' he says, 'gorgeously gowned and got up to captivate. She tripped along lightly, only the rustling of her silk garments being audible. As she advanced she looked around the saloon for a seat, but accidentally, I suppose, made a bee line for Mr Rhodes's table. She appeared quite overcome with surprise when she did see him and exclaimed, "Oh! How do you do? . . ." She nervously placed her hand on the back of a chair at our table and said, "Is this chair engaged? May I have it? . . ." Of course she occupied the chair for the rest of the voyage.'

For all its happy detail, this account is open to question. So, for that matter, are the stories which the secretary tells of Princess Radziwill hounding Rhodes throughout the voyage: fainting in his arms and forcing him to flee to the captain's deck to escape her. Others on board the *Scot* were unaware of this blatant pursuit. What is certain is that, by the time they reached Cape Town, Rhodes had got to know the Princess very well.

At first he was friendly towards her. She stayed at the Mount Nelson Hotel and Rhodes frequently invited her to Groote Schuur. Often she would arrive at the house uninvited. Then things began to go wrong. Rhodes got tired of her attentions and, it is said, would have a horse saddled ready to make a quick escape whenever she appeared at his gate. He pleaded with his friends not to leave him alone with her. More alarming was a report by one of his secretaries

that some important papers had disappeared from his office after one
of the Princess's visits.

The turn of events in South Africa eventually freed Rhodes from
the troublesome Princess. In October 1899 the Anglo-Boer war
broke out. The war climaxed the continuing conflict between
President Kruger and the uitlanders in the Transvaal. Although
Rhodes had wisely kept aloof from the final stages of this quarrel, he
was determined to play a part in the war. His thoughts turned
immediately to Kimberley—the cradle of his fortunes. Kimberley
represented his great beginnings and De Beers Consolidated Mines
was still the main source of his wealth. When it was rumoured that
the town would be besieged, Rhodes left Cape Town and made
straight for Kimberley. He arrived there two days before the besieg-
ing force and was shut up in the town for four months.

While he was away, the Princess busied herself cultivating his
political opponents in the Cape, claiming that she was acting as his
political agent. Rumours were spread that Rhodes intended to marry
her. It was said that he had visited her secretly in her room at the
Mount Nelson.

Rhodes was furious when he returned and heard what she had
been up to. The Princess, herself, admitted that they had several
rows about her political meddling. Unfortunately, she was in no
position to argue. She was up to her neck in debt. The hotel was
threatening to turn her out. A friend told Rhodes about this and he
offered to pay the Princess's debts if she would leave South Africa.
'I paid her bills and she left the country,' he was to say, 'but she
came back again.'

She came back, after a few months in London, to continue her
harassment of Rhodes. There were more rows. The final break came
when she refused to hand over to Rhodes certain papers. These may
have been the papers that were stolen earlier from his office. Her
refusal to part with the papers was the last straw. Rhodes would have
nothing more to do with her. During the months that followed he
was away for most of the time, travelling in Rhodesia and Europe.
But the Princess did not let up. She started a pro-Rhodes newspaper.
She hinted that Rhodes was prepared to back her. Promissory notes,
in Princess Radziwill's favour and bearing Rhodes's endorsement,
began to circulate in Cape Town. Rhodes was wired but denied all
knowledge of them. The banks and the money-lenders were, for the
most part, wary of accepting them.

This did not deter the Princess. She sank deeper into debt and became hopelessly entangled in her own machinations. By this time Rhodes was in Europe. He left it to his agents in South Africa to tackle Catherine Radziwill. One of them honoured a promissory note for £2,000, supposed to have been signed by Rhodes. Then he demanded repayment. The Princess could not pay and faced imprisonment.

But she was given a chance of evading the law. She was visited by Rhodes's friends, by a lawyer and by a Government official; they all promised to release her from her financial responsibilities if she would hand over her 'papers'. Exactly what these papers contained has never been disclosed. It seems likely that they were copies of the notorious 'missing telegrams' connected with the Jameson Raid. They never came to light because the Princess refused all the offers made to her. The law began to take its course.

As Rhodes's signature appeared on the promissory note, his presence at any legal proceedings was essential. He was summoned to appear in the Supreme Court in Cape Town. Ill as he was at the time—his heart was failing and he was not expected to live long—he insisted on returning to South Africa. His friends did their best to dissuade him. They pointed out that the summer heat at the Cape was stifling; that he was in no condition to face the strain; that his evidence could be given by affidavit. But he would not listen to them. He was determined to face the Princess in court.

The case was heard on 6 February 1902. The court was packed. News of the Princess's 'papers' had leaked out and it was naturally expected that these papers would reveal, not political, but romantic secrets. Rhodes was well known as a 'woman-hater' and a confrontation between him and the Princess promised to be lively. But there was no confrontation. The Princess did not appear: she pleaded sickness. Rhodes, looking ill and panting for breath, gave his evidence and disclaimed all knowledge of the promissory note. The judge found that the signature was 'clearly an absolute forgery'.

And there the matter might have ended. Despite the judge's verdict of forgery, no action was taken against the Princess. It was she who took the next step. For some inexplicable reason, she had a summons served on Rhodes, suing him for payment of the £2,000 bill. 'Damn that woman!' Rhodes exploded when his secretary handed him the summons. 'Can't she leave me alone?' This time the Princess had gone too far. Sending for his lawyer and a magistrate,

Rhodes drew up an affidavit accusing Catherine Radziwill of forgery.

On 25 February 1902, Princess Radziwill was formally charged with forgery and uttering a forged document. During the preliminary hearing of the case, she came face to face with Rhodes for the last time. Too ill to attend court, Rhodes had his evidence taken by the magistrate at Groote Schuur. The Princess was present with her attorney. 'He gave his evidence . . .', wrote a journalist, 'from his seat on the couch, and never once looked at the Princess as she sat with her solicitor at the back of the little circle of people in front of the Magistrate's desk. But the Princess never took her eyes off him. He was dressed in a grey jacket coat, white flannel trousers, and black boots.' Rhodes's evidence was a prepared statement to the effect that he had never signed any of the documents then before the court. The effort of appearing before the magistrate was almost too much for him. He was reported to be looking very ill and coughing badly.

'Almost immediately after he had signed his statement,' says the journalist, 'he left the room by the door through which he had entered. . . . But before he went from the room the Princess was asked if she had any question to put.

' "No," said her solicitor, speaking for her.

' "But I will speak," insisted the Princess in a dramatic stage whisper, as she half rose.

' "No, no; sit down," peremptorily ordered the solicitor and the lady obeyed him.'

What the Princess wanted to say—what she could possibly have said—remains a mystery. The relationship between Princess Radziwill and Cecil Rhodes is only slightly less inexplicable than that between Baron von Veltheim and Barney Barnato's nephew.

Rhodes was so obviously unattracted to women that the Princess's contention that he was in love with her is difficult to accept. She was undoubtedly drawn to him, she might even have been in love with him, but he certainly did not reciprocate her feelings. That she had some hold over him and that that hold was political is fairly clear. Rhodes's one concern was to get her to return the 'papers' she held and in this he was aided by lawyers and Government officials. Only when she refused to hand over the papers did Rhodes move against her. What those papers contained and how she got hold of them, the Princess never revealed. It is probable that they were connected with the Jameson Raid, they might have been the 'missing telegrams'; it is probable that she stole them. Like von Veltheim, she preferred not to tell the truth in court; she was prepared to be punished for forgery rather than admit to blackmail.

And punished she was. At her trial in the Criminal Sessions of the Supreme Court at Cape Town, she was found guilty on twenty-four counts of fraud and forgery. It was shown that the signature on all the notes she had issued had been traced from a signed photograph of Rhodes. She was sentenced to two years' imprisonment in a 'House of Correction'.

Rhodes never knew the outcome of her trial. At three minutes to six on the evening of 26 March, 1902, he died. The black-bordered newspapers reporting his death carried an account of the last preliminary hearing in the trial of Princess Radziwill. Sentence was passed on her in the Supreme Court a month later.

'So it came to pass,' wrote W. T. Stead, 'that he who had never harmed a woman in his life met his death in clearing his name from the aspersions of a woman whom, out of sheer good-heartedness, he befriended in time of need.'

This was a rather obvious exaggeration but it was believed by a great many people.

[6]

Rhodes was forty-eight years old when he died. Even without the intervention of Princess Radziwill, he would not have lived much longer. All his life he had suffered from a weak heart; he had always expected to die young. He was haunted by the thought of not completing what he had set out to do. 'Everything in the world is too short,' he once said. 'Life and fame and achievement, everything is too short.' And, on another occasion, he declared: 'From the cradle to the grave, what is it? Three days at the seaside.' This was a recurring theme with Rhodes. It was probably more responsible for his early death than was Princess Radziwill.

He was always working against time. Into his brief forty-eight years he crammed the work of several lifetimes. He had arrived in South Africa, at the age of seventeen, an unknown sickly boy. By the time he was thirty-seven, he had educated himself; built up one of the largest financial empires in the world; organised the conquest of a country; become Prime Minister of the Cape. But this was not enough. It was his dream to master first Africa, then the world; and his dream would allow him no rest. He pushed on with his plans for

linking the Cape to Cairo; for furthering British interests throughout the continent of Africa. His life became a series of crises; financial, political and military. Triumphs were followed by set-backs and
set-backs by disasters. Few robust, healthy men could have stood
the strain of such a life. And Rhodes was far from healthy, certainly
not robust. He suffered repeated heart-attacks, he contracted malaria,
he met with serious accidents. For weeks he was laid up, ravaged by
disease or crippled by physical injuries, but he gave himself no time
to recuperate. Unlike as were he and Barney Barnato in most
respects, they shared the same fatal tendency to drive themselves to
extremes. Their restless spirits had made them what they were; it
was their inability to relax that killed them.

Rhodes was a doomed man long before he returned to South
Africa in that hot summer of 1902. Doctors had warned him that the
end was near. Friends who met him in England were shocked at his
altered appearance. 'His face was bloated, almost swollen,' said one,
'and he was livid with a purple tinge in his face, and I realized that
he was very ill indeed.' The voyage to the Cape had not helped him.
Not only did he catch a severe cold but one night, when sleeping on a
writing table in his cabin in an attempt to catch a cool breeze, he fell
and badly bruised his nose and shoulder. 'It is a marvel he was not
killed,' says his secretary.

There was little hope of his surviving the dreadful heat of the
Cape. Every day he paced the great rooms of Groote Schuur trying
to find relief from the stifling atmosphere. With his shirt unbut-
toned, his hair matted and the sweat pouring from his brow, he
would slump, panting, on to a couch in the darkened drawing-room,
soon to get up again and drag himself to an open window upstairs in
the hope of finding a breath of air. But there was no breeze. The heat
pressed down on the huge white house, the garden shimmered in the
scorching sun and the trees were still. Only at night could he hope for
a brief respite. Then he would drive to his little cottage on the beach-
front at Muizenberg, on the outskirts of Cape Town. He would sit
on the verandah of the cottage, listening to the crash of the breaking
surf, and sometimes he would find relief in a breath of cool sea
air.

He died at the Muizenberg cottage. For three days he lay there
more or less unconscious, watched over by his friends. Dr Jameson
hardly left his side. His favourite secretary slept on a stretcher in his
room. When the end came, Jameson went on to the verandah of the

cottage and told the waiting crowd. His last words, said Jameson, were: 'So little done, so much to do.'

No more appropriate words could have matched his end. They were perhaps a little too appropriate. Jameson was quoting a prepared statement; those who stood at Rhodes's bedside gave a different account. They say that immediately before he died, he stirred and muttered the names of his friends. His last words were addressed to his secretary. He said: 'Turn me over, Jack.' He had lived as The Colossus, but he died a human being.

There were more words yet, however. These were the words of his will. He had made several wills during his lifetime; the final version was drawn up in London on 1 July 1899. This important document was witnessed by his friend, Sir Charles Metcalfe, one of his private secretaries, Philip Jourdan, and Arthur Sawyer, his waiter at the Burlington Hotel. Codicils were added in 1901 and 1902, but essentially the original provisions remained unaltered.

It is estimated that, at the time of his death, Cecil Rhodes's estate was worth over £6,000,000. His holdings in De Beers Consolidated Mines accounted for £2,500,000 of this. He made a number of minor bequests and stipulations. Groote Schuur and its furniture he left in trust as a residence for future Prime Ministers of a 'Federal Government of the States of South Africa'. To his old college at Oxford, Oriel College, he bequeathed £100,000 for the erection of new buildings and the future benefit of the Fellows of the college. He made provisions for his family and for the upkeep of his various properties and left annuities of £100 each for two of his servants. His secretaries were provided for in separate instructions to his executors.

The most important provision of his will, however, was for the establishment of the famous Rhodes Scholarships. By means of these scholarships he hoped to ensure that a never-ending stream of clever, healthy young men, educated at Oxford, would continue his work of spreading Anglo-Saxon ideals and influence throughout the world. They were to be chosen for their literary and scholastic achievements, their fondness of outdoor sports, their qualities of leadership and moral force of character. They were to be the paragons of manhood whom Rhodes so admired but could never emulate. This was the outcome of his youthful dream of a Secret Society which would one day transform the world. 'I contend that we are the first race in the world', he had written in 1877, 'and that the more of the world we inhabit, the better it is for the human race.'

His life had been dedicated to this ideal: he died with his faith intact.

He made another stipulation in his will. 'I admire the grandeur and loneliness of the Matoppos in Rhodesia', he stated, 'and therefore I desire to be buried in the Matoppos on the hill which I used to visit and which I called the "View of the World" in a square to be cut in the rock on top of the hill covered with a plain brass plate with these words thereon—"Here lie the remains of Cecil John Rhodes".' His instructions were faithfully observed.

At an impressive ceremony, on 9 April 1902, Rhodes was buried in the Rhodesian hills. The slopes of the 'View of the World' were lined by thousands of Matabele. As Rhodes's coffin reached the top of the hill, a great shout rang out: 'Bayete!'—the Matabele salute to a great chief.

CHAPTER TWENTY ONE

'ONE THAT LOVED HIS FELLOW MEN'

IN March 1888, four men had spent an entire night arguing about the terms of the Trust Deed of De Beers Consolidated Mines. In a bleak, corrugated iron Kimberley cottage, Cecil Rhodes, Alfred Beit, Barney Barnato and Woolf Joel had guaranteed extensive political powers to one of the largest financial organisations in the world.

If the setting was incongruous, the relative youthfulness of the financiers was extraordinary. Of the four, Barnato was the eldest. He was then thirty-five. Rhodes was a year younger, Beit had only just turned thirty-five, Woolf Joel was still in his twenties. Yet, young as they were, none of them was destined to live much longer. By 1902—a mere fourteen years later—three of them were dead. Barney Barnato had killed himself at the age of forty-four, possibly worried by the intrigues of Baron von Veltheim; Woolf Joel was shot by von Veltheim when he was in his mid-thirties; Cecil Rhodes had died at forty-eight, trying to free himself from the machinations of Princess Radziwill. The fourth actor in that all-night drama, Alfred Beit, lived slightly longer. He nevertheless died young and his death was not peaceful. Those fatal contributory causes—worry, overwork and the meddling of a foreign adventurer—all played their part. But at least Beit was not to blame for his misfortunes; he did not seek the troubles that beset him.

'All that Beit wanted', Rhodes once said, 'was to be rich enough to give his mother £1000 a year.' And this, strange as it may seem, was probably true. There had been times, during his early Kimberley career, when Beit would willingly have retired. Once he had made a large enough fortune to enable him to live comfortably for the rest of his life, he wanted nothing more. He had no financial or political ambitions. What kept him at work was his loyalty to his partners— first to Porges and Wernher, then to Rhodes—and his undoubted financial genius. Like any gifted person, he was unable to ignore his own abilities and his mastery of finance drew him, almost unwillingly, into the world of big business. But it was his sense of duty rather than greed that kept him there. As long as his gifts could be used to benefit his firm or his friend he felt honour bound to exploit them. He was rarely heard to express a desire of his own.

In 1889, shortly after the amalgamation of the diamond mines, Beit went to live in London. He took rooms in Ryder Street and worked, as managing director, in the City office of Wernher, Beit & Co. By that time his firm's extensive interests on the Witwatersrand were expanding under the able direction of Hermann Eckstein who, assisted by J. B. Taylor, was recognised as one of the most influential men in the gold mining industry. Both Eckstein and Taylor were Beit's appointees and there can be little doubt that Beit was largely responsible for their initial success. Not only had his intuitive genius placed him in the forefront of the Transvaal gold rush, but his sound business principles had ensured the firm's prosperity.

The example set by Beit's firm—soon to become well-known as the Corner House—was followed by other mining companies; it contributed greatly to the stability of the gold mining industry. Beit, in fact, is widely acknowledged as the 'architect' of the Witwatersrand. 'Beit's was the master mind in making a success of the gold mining industry on the Rand,' claimed J. B. Taylor. 'From the first he resolved that the mines under his firm's control were not to be run for sharemaking and marketing purposes. For in no instance did the firm issue a prospectus. The working capital was always found by the firm and the companies financed until they became dividend paying. The shareholders were informed monthly of everything that happened on the mines—nothing was hidden. It was a complete revolution in mining history.'

Such was Beit's reputation for honesty that, inevitably, others traded on it. Rich, kind-hearted and easy-going, he was, in many ways, a natural victim for the unscrupulous. Often he was taken advantage of; rarely did he try to defend himself. There were times, however, when his lenient attitude involved him in endless trouble. One such episode came to a head shortly after his arrival in London. It resulted from the dubious activities of his cousin William Lippert.

[2]

It had been to join his cousins, William and Edouard Lippert that Beit had first come to South Africa. Although his association with the Lippert firm had not lasted long, the family ties had remained firm. Beit had a strong sense of family and it was this that his cousin William appears to have relied upon. And so he could—had he not been so reckless.

The trouble started with William Lippert's association with the Union Bank of Cape Town. This was a well-known private bank which had been in existence for forty-three years; some of the leading citizens of the Cape Colony were listed among its shareholders. Despite this respectable front, the bank was not as solid as it at first appeared. An uncritical directorate relied entirely on the manager who, in turn, was apt to be a little too liberal with the credit he allowed. The books were rarely, if ever, audited and none of the distinguished shareholders was aware of the bank's accumulating losses. This was a particularly dangerous state of affairs as the bank was not a limited liability company and the shareholders could be held fully responsible for its debts. Like most South African banks, the Union was very much involved in mining speculation and when a slump occurred on the Rand it found itself in difficulties.

William Lippert seems to have been one of the few who were aware of the sorry state of the Union Bank. In an effort to tide it over the crisis he issued certain securities, including a bill signed by himself and to which the forged signature of Alfred Beit was added. When Beit was presented with this bill, he paid up without a word. Lippert, encouraged by his cousin's silence, then issued another bill and again Beit honoured the forgery. However there was a limit even to Alfred Beit's indulgence. That limit was reached in 1889 when he was presented with a forged bill for the third time. Having already paid out £150,000, Beit was forced to take action against William Lippert. He telegraphed his agent in Kimberley, exposing the forgery. In the investigation which followed more than Lippert's misdeeds were revealed. The position of the bank was shown to be hopeless and the shareholders, large and small, were held responsible. A great many investors were ruined by the crash.

William Lippert was among the worst hit. His losses were estimated at over £400,000. Unable to face the consequences, he left South Africa and went into hiding. For five years he lived in the United States, under a false name, eking out an existence by selling books. Eventually he decided to end his miserable exile and returned to stand trial in the Cape Supreme Court. In 1895 he appeared before the Chief Justice of the Cape Colony; Alfred Beit was called to give evidence against him.

The trial was a gruelling experience for Beit. When giving evidence, his nervousness was painfully obvious. As was to be expected, he was roundly castigated for not exposing the forgeries earlier. Taxed about this by the judge, he replied pathetically: 'I thought of only one thing—to save the family.'

Lippert was found guilty and sentenced to seven years' hard labour. His sentence was regarded as lenient. 'For the last five years', the judge allowed, 'you have been a fugitive on the face of the earth, and I fully believe and concur in what your counsel says that the mental suffering you must have borne during that period must have been almost as great as if you had been undergoing your punishment here. . . .'

The money Beit had lost was later repaid by the Lippert family. It could have been small consolation for the misery he had endured. His misguided loyalty had resulted in the public disgrace that he had tried so desperately to avoid. He was openly blamed for supporting his cousin. Soon he was to be similarly blamed for supporting his friend.

[3]

'Rhodes could never have achieved what he did at Kimberley nor in Rhodesia without Beit,' says Lionel Phillips. This was truer than most people realised. The amalgamation of the diamond mines owed as much to Alfred Beit's financial acumen as it did to Cecil Rhodes's clever strategy. The one would have been hamstrung without the other. No less important was the backing which Beit gave to Rhodes's Chartered Company. Both financially and morally, 'little Alfred' contributed as much to the opening up of Rhodesia as he had to the launching of De Beers Consolidated Mines.

Although Beit had no interest in politics, he was always prepared to follow Rhodes's lead. He readily became a director of the Chartered Company and his financial support did much to encourage confidence in the venture. In 1891, a year after the occupation of Mashonaland, he accepted Rhodes's invitation to visit the territory. Rhodes needed his assistance. Not only was Mashonaland's mineral potential in doubt but the Chartered Company was experiencing some difficulty in exercising its monopoly over the territory.

The chief difficulty arose from a concession which William Lippert's brother, Edouard, had obtained from Lobengula. By claiming that the Rudd Concession applied only to mineral rights,

Edouard Lippert had persuaded the Matabele chief to grant him certain land and settlement rights. The motive for Lippert's interference is somewhat uncertain. He is said to have been activated by his jealousy of his cousin—for some reason he had quarrelled with Beit—and his known dislike of Rhodes. There was also a suspicion that he was being backed by the Transvaal Government. In any case, Rhodes was determined to get rid of him and apparently thought that Beit, despite the family quarrel, could help in buying Lippert off.

Beit sailed for South Africa in April 1891. His departure from London was hardly noticed. From the very outset, his expedition to Mashonaland was completely overshadowed by the publicity given to his travelling companion, Lord Randolph Churchill. Like Beit, Churchill was also visiting the Chartered Company's territories at the invitation of Rhodes. The two expeditions had been arranged simultaneously, but it was Lord Randolph's that attracted attention. Accompanied by a retinue of servants and several overloaded wagons, Churchill travelled through South Africa in a grand manner. His progress—reported in a series of fault-finding, controversial letters to the London *Daily Graphic*—kept the public amused for months. Few were aware of the modest, retiring Beit who followed in his wake.

The Mashonaland expedition revealed an unsuspected side of Beit's personality. Far from resenting the hardships he was forced to endure, he seemed to enjoy bumping along the primitive roads, sleeping in his wagon and eating beside a camp fire. His life, both in Kimberley and London, was meticulously ordered and the freedom of the veld undoubtedly had an invigorating effect upon him. It broke down his customary reserve. Surrounded by the Chartered Company's police at a camp fire concert, he became positively envious of their way of life.

'What's the use of being a millionaire?' he asked. 'None. What good is money? None. A trooper's life for me—no cares, no troubles, and all the world before you, no life like a trooper's!' When it was suggested that he might be happier as an officer, he disagreed. 'No,' he declared, 'officers have ambition; he has ambition and wants to get on. A trooper's is the life for me.'

As transport officer for his wagons, he employed Percy Fitzpatrick, a young South African of Irish descent. Fitzpatrick also had the unenviable task of supervising part of Lord Randolph Churchill's expedition. He found a marked contrast between his two employers. Where Churchill was aloof, high-handed and critical, Beit was friendly, warm-hearted and sympathetic. Churchill's only interests in Mashonaland were hunting and looking for gold. Beit was greatly concerned with the plight of the struggling settlers and supplied Fitzpatrick with 'a considerable sum in cash' to help any pioneers they encountered.

They met a number of stranded travellers, including Mr and Mrs Edouard Lippert. The Lipperts had run out of provisions and were only too pleased to be rescued by Beit. Whatever hostility had existed between them melted during the three days they spent together. Fitzpatrick was to remember this time as a delightful picnic. Unfortunately the *rapprochement* was short-lived; but while it lasted it did some good. It may well have contributed to Lippert's later decision to sell his controversial concession to Rhodes.

Fitzpatrick found Beit a charming and amusing travelling companion. He was full of anecdotes about his experiences in Kimberley and on the Witwatersrand. Much that is known about Beit stems from his Mashonaland expedition, for he spoke more freely to Fitzpatrick than he did to others. They got on famously. Both had a sharp sense of humour and they spent much of their time laughing. Not all of Beit's jokes were intentional, however. Fitzpatrick—then a hard-up transport officer—could not help chuckling when, in all seriousness, Beit closed a conversation by saying: 'Yes, it is the first million that takes a lot of making.' 'But,' says Fitzpatrick, 'his heavy sigh broke off into bubbles of laughter as he realised the humour of it from my point of view.'

Beit was to show his appreciation of Fitzpatrick's position in a more positive way. When the expedition was over, he took his transport officer into his employ on the Witwatersrand. For Fitzpatrick, this was the beginning of a brilliant career.

Beit set his mining experts to work examining Mashonaland's mineral potential. Although he was by no means optimistic about their findings, he expressed every confidence in the country. 'He does not disguise from himself that the Company's difficulties are not yet at an end,' it was reported. 'At the same time he is firmly convinced that the Company possesses a very fine country, not only from an agricultural but from a mineral point of view. Mr Beit's watchword in connection with Mashonaland is PATIENCE.'

Unhappily, Beit's watchword did not appeal to Rhodes. And

when Rhodes's patience was exhausted, he carried Beit along with him. Just as Beit had assisted in the amalgamation of the diamond mines and the launching of the Chartered Company, so he supported the disastrous Jameson Raid. It was his money as much as Rhodes's that supplied the uitlanders with arms before the Raid; and it was he who helped Rhodes pay the fines imposed on the leaders of the Reform Committee.

Beit, it is said, did not expect the Raid to result in bloodshed. Although he financed the arming of the uitlanders, he was certain that the Boers would yield before a shot was fired. Jameson's impetuosity and the firm stand taken by Kruger proved him wrong. He was completely shattered by the failure of the Raid. What troubled him most was the arrest of the Reform Committee. One of the leaders of the Committee was his employee, Lionel Phillips (who had worked for J. B. Robinson in Kimberley) and the thought that he was partly responsible for Phillips's imprisonment weighed heavily with Beit.

At the time of the Raid, Beit was staying with Rhodes at Groote Schuur. Lady Sarah Wilson, who met him there shortly after the receipt of the news of Jameson's capture, says that he appeared 'quite crushed at the turn events had taken—not so much on account of his own business affairs, which must have been in a critical state, as in regard to the fate of Mr Lionel Phillips'. Unlike Rhodes, Beit was unable to shake off his feeling of guilt. He managed to pull himself together sufficiently to return to London and attend to the affairs of the Chartered Company but he was still sick with worry. Lionel Phillips's wife saw him in London and was shocked by his pitiful appearance. 'I never saw any one look so utterly, hopelessly wretched,' she declared. 'I can honestly say that I believe the sufferers and victims in prison did not suffer as much for themselves as he did for them.' Beit had few enemies: with Rhodes for a friend he did not need them.

There was worse to come. At the Committee of Inquiry hearings into the Raid, Beit came in for more than his share of the blame. He was viciously attacked by Henry Labouchere, the Radical M.P. who edited the periodical *Truth*. In a speech in the House of Commons, in a series of articles in *Truth* and in a caustic letter to the French newspaper *Le Gaulois*, Labouchere denounced Beit as a sinister capitalist who had engineered the Raid to benefit his own pocket.

'Mr Beit . . .', Labouchere stormed, 'and several other gentlemen of that kidney met together in the city. They considered that, although they might eventually gain by laying hold of this portion of the Transvaal, undoubtedly the first effect of a raid . . . would be a heavy fall in all the shares of the South African market. They therefore confederated together and employed a gentleman to sell for them what was called in the city an enormous "bear" in these securities, in order that they might gain.'

There was not a word of truth in this. Subsequently Labouchere was obliged to withdraw his accusations and reluctantly admit Beit's integrity. But the harm was done. For years Alfred Beit was regarded by the ill-informed as an 'ogre and business man, who sacrificed everything to money-making'.

The Select Committee exonerated him from Labouchere's insinuations. They admitted that there was not the slightest evidence of stock-jobbing. But on the main issue they found that: 'Mr Beit played a prominent part in the negotiations with the Reform Union: he contributed large sums of money to the revolutionary movement, and must share full responsibility for the consequences.'

Once again Beit was made to suffer in public for his misguided loyalty.

[4]

In 1894, shortly before the Jameson Raid, Beit had moved into his Park Lane mansion. He had bought the site for this house from the Duke of Westminster after some hard bargaining. The Duke, it is said, had been somewhat dubious about selling out to the South African *nouveaux riches* and had been reluctant to come to terms. So hesitant was he, in fact, that on the eve of signing the agreement he had sent an urgent note to Beit asking for an assurance that at least £10,000 would be spent on the house. Beit, tongue in cheek, replied that he intended spending that much on the stables alone. (Barney Barnato started building shortly afterwards and this rejoinder is sometimes attributed to him, but it seems to belong to Beit.)

The house—26 Park Lane—although by no means as ornate as Barnato's, was of little architectural merit. What slight distinction it had came from the fact that, in contrast to its imposing neighbours, it was a relatively low, two-storeyed building. One of Beit's acquaintances described it as 'a cross between a glorified bungalow and a dwarf Gothic country mansion'. For all that, it was spacious and extremely comfortable. Beit's favourite retreat seems to have been the luxuriant winter garden which opened off the large drawing room. Visitors to the house were invariably impressed by this novel Germanic feature. 'We were seated in the room which was at once a sort of rockery and palm garden,' says Frank Harris who interviewed Beit there; 'a room of brown rocks and green ferns and tesselated pavement—an abode of grateful dim coolness and shuttered silence made noticeable, as it were framed off, by the vague hum of the outside world.' After the dust, heat and noise of those long years in Kimberley, this must have been a haven indeed.

Despite his lack of personal ambition, Beit undoubtedly was attracted to luxury and the things that his immense riches could buy. He was at home in comfortable surroundings; he loved good food and wine; he had an instinctive appreciation of art and music. He came to be regarded as 'perhaps the most cultured of all the Rand magnates'.

At Covent Garden he kept a box permanently reserved for himself and his friends and could speak knowledgeably about opera and the theatre. His study was lined with the works of German classical writers and contemporary English historians. George Eliot was his favourite novelist and he also admired Thackeray and Trollope; but, with an independence of judgement, he never succumbed to Dickens.

His greatest pride was his collection of art treasures. Hunting for paintings and bronzes was one of the few private passions he indulged to the full. He would travel great distances to track down a work of art and he was never too tired to interview the dealers who flocked to his house or to seek an artist out in his studio. Opinion differs about his aesthetic appreciation. Frank Harris found it poor. 'He had a real love and understanding of music,' says Harris; 'and he admired pictures and bronzes, too, though he was anything but a good judge of them, and his taste here illustrates a side of his character which is not generally appreciated. Mr Beit gave an enormous sum for all the Dudley Murillos, simply because Murillo's type of sentimental Southern beauty appealed to him intensely. He thought a certain long-limbed angel with uplifted eyes a sort of perfect ideal or model of the beautiful, and was astounded to find that one rather disliked it. At bottom Mr Beit was a sentimentalist, and did not count or reckon when his feelings were really touched.' Certainly sentiment would not have played much part in Harris's judgement and his criticism might well have been justified. However, others were not so conscious of Beit's weakness. His biographer, Seymour Fort, not only praises the Beit collection but says that he never purchased an important picture without taking advice. If he was influenced by sentiment, it was not cheap sentiment. Besides his Murillos, his collection included paintings by Rembrandt, Hals, Hoppner and Reynolds. 'Beit's artistic tastes are above the average,' remarked Sir Lewis Michell.

His social pretensions were as modest as his political ambitions. One of the richest bachelors in London, he was much sought after but he had little taste for social life. He preferred to entertain a few select friends rather than attend large social functions. The trivialities of London society bored him. In a small gathering he could be a friendly, attentive host but he was poor value as a guest. An intelligent and sympathetic listener, he was not a great conversationalist. Most of his friends were business associates or members of the Anglo-German community; occasionally he would entertain a social celebrity or a politician—he was popular with the Churchill family— but outside his own set he preferred the company of actors and artists.

His interest in women appears to have been as slight as was Rhodes's. Despite the allure of his money, no woman outside his family ever managed to get close to him. But unlike Rhodes he did not cultivate young men nor did he gain the reputation of being a woman-hater. His emotions were tuned to a much lower key. 'He enjoyed the society of women very much as a moderate drinker enjoys an occasional glass of good port,' says Seymour Fort, 'that is to say, limited in amount and frequency.' If Alfred Beit had a sex life, it was very discreet.

He was perhaps conscious of his own plainness. Diminutive and bald from an early age, he was far from handsome: his gentle benign looks lacked any semblance of sexual attractiveness. In Kimberley he had made a point of choosing tall, statuesque women as dancing partners and, if this was an indication of his sexual preference, his own dumpiness must have caused him embarrassment. 'Mr Beit?' exclaimed an actress, on meeting him for the first time. 'Beit did you say . . .? Looks more like a nibble to me.' For one as shy and sensitive as Alfred Beit, an immense fortune could not offset the possibility of ridicule.

The house in Park Lane was shared by two friends. One was his cousin, Franz Voelklein, who had once represented the Lippert firm in Port Elizabeth and who was now Beit's constant companion and private secretary. The other was his fox terrier, Jackie, to whom he was devoted. There appears to have been a series of terriers named Jackie in Beit's life, all of whom he cherished.

J. B. Taylor tells a story of a journey which he and Hermann Eckstein once made with Beit, travelling from Kimberley to Pretoria. Beit was nursing a terrier puppy which he was taking as a present for Taylor's sister. About twenty miles from Pretoria the axle of the coach snapped and the coach overturned. It was midnight. Taylor scrambled out of the coach to find the driver lying stunned by the roadside and all else a deathly silence. 'I called out to Beit and Eckstein, asking if they were hurt,' says Taylor. 'The only answer I got was: "Where is Jackie?" (the pup). One by one I helped the passengers out of the coach till only Beit remained. Everyone had trodden on him trying to get out, but in his search for Jackie he felt nothing.' Eventually Jackie was discovered, unhurt, sleeping under a pile of rugs and cushions; only then did Beit crawl from the coach and enquire about the other passengers.

His affection for his dogs was second only to his love for his family. Nothing in his life ever equalled his devotion to his mother. When he built 26 Park Lane, he had a similar house erected in Hamburg for his mother. He took as much interest in planning his mother's house as he did his own and went over to Hamburg to supervise personally the laying out of the extensive garden 'on modern landscape lines'. In Hamburg he was always happy and he visited his mother whenever he could. When they were apart they wrote to each other constantly; letters full of family gossip. Beit's biographer, examining these letters, was surprised to find that they contained little about his life in South Africa or his business concerns. They simply reflected 'his profound affection for his home and everything connected with it'.

Throughout his life he retained his love for Germany. One of his few known ambitions was to promote an Anglo-German *entente*. It was with this in mind that he later acquired a controlling interest in a newspaper called the *Anglo-German Courier*. But his home was England and the longer he lived there the more British in outlook he became. The seeds sown by Cecil Rhodes fell on very fertile ground. 'Beit,' says Lionel Phillips, 'although a German by birth, was a keen imperialist.'

At Queen Victoria's Jubilee celebrations, in 1897, no house was more lavishly festooned with Union Jacks than 26 Park Lane. So carried away was Beit on this occasion that his habitual reserve broke down and he attended the Duchess of Devonshire's famous fancy dress ball, dressed in velvet and lace, as a Stadhouder of Holland. The following year he became a naturalised British subject.

To live up to his status as an English gentleman, he started negotiations for a country house. His doctor had advised him to find a quiet retreat away from London and, in 1902, he purchased Tewin Water, a seven hundred acre estate near Welwyn in Hertfordshire. The Regency style house had belonged to Lord Cowper and Beit took it over as a 'going concern'—complete with servants, furniture, plate, linen, even fishing rods and tackle. When the time came for him to take possession, however, Beit was overcome by shyness. 'In this emergency', says Sir Lewis Michell, 'he asked the late Earl Grey and myself to accompany him there as a protection against the unknown. We accordingly all three went to Tewin Water together and spent a delightful weekend, Beit being in the highest of spirits over his new toy.' But, as he was later to discover, it was an expensive toy, even for a millionaire. When a visitor remarked on how nice it must be for him to have his own milk, butter and eggs, his reply was tart. 'Yes', he said, 'but don't forget eggs cost me 6d. each, milk 1 shilling a pint and butter anything up to 10 shillings a pound.'

His servants, like all his employees, soon came to adore him. Nobody could work long for Alfred Beit without recognising his good qualities. In the mining world he was looked upon as an exceptional employer. 'He always takes a deep interest in the private life of his clerks and engineers, by whom he is regarded with the utmost affection,' it was said. 'When he speaks to them he is always as one of themselves. His manner is always simple, courteous, and sympathetic, and he likes to know that they have some definite interest outside the business; that their lives are not entirely consumed in the contemplation of ledgers and plans.'

[5]

Rhodes had died shortly before Beit moved to Tewin Water. Those who were with Beit when he learned of his friend's death, were never to forget his grief. He said little but his suffering was painfully obvious. When he did bring himself to speak, he spoke only of the

importance of continuing Rhodes's work. This was to be his chief
concern until his own death four years later.

In August 1902, Beit sailed for South Africa on the fateful
s.s. *Scot*. Accompanied by Dr Jameson, he intended to investigate
Rhodes's many concerns. Both Jameson and Beit were trustees of
Rhodes's will, but it was their strong sense of friendship which
inspired their work. Jameson regarded himself as Rhodes's political
heir and Beit was prepared to back him financially.

One of their first concerns was to establish appropriate memorials
to Rhodes. Memorial Committees had been set up in Cape Town and
Kimberley and Beit spent hours straightening out their problems
and coordinating their activities. In Kimberley he was accompanied
by Sir Lewis Michell who found the contrast in his business
methods fascinating. The Kimberley Memorial Committee was in a
dilemma. Having spent all the money collected on an equestrian
statue of Rhodes, they had nowhere to place it. De Beers had
promised them a site but they considered this highly unsuitable
and preferred a spot closer to the cottage in which Rhodes and
Jameson had lived. 'Beit', says Michell, 'replied simply that "he
would look into it", a formula which often means nothing at all. But
the next morning he went with me to the place and approved of it,
went to the owners and bought it at a high figure, and then handed it
over to the Committee. Yet an hour later he had in my presence an
amusing altercation with a cab driver over a shilling!'

After visiting Cape Town, Johannesburg and Kimberley,
Jameson and Beit went on to Rhodesia. This was Beit's particular
sphere of interest. Since his visit in 1891, he had been actively concerned with the territory. The transport problems of the remote
region had spurred him to sponsor the extension of the railway line
from Kimberley to Bulawayo and to act as chairman of a company
which aimed to provide a railway link between Rhodesia and Beira,
thus providing the territory with an outlet to the sea. Although he
had been obliged to resign his directorship of the Chartered Company after the Jameson Raid he had been reinstated on the board
shortly before he arrived in South Africa and could now act with
authority. Both Jameson and Beit were kept extremely busy. Lady
Sarah Wilson, who met them in Salisbury, was struck by the way in
which the pioneers turned to them now that Rhodes was no longer
there. Rhodes's 'faithful lieutenants', she says, 'were doing their best
to replace him, and the role of Jameson, apparently, was to make the

necessary speeches, that of Beit to write the equally important cheques'.

The strain of the long arduous journeys and the constant pressure of work proved too much for Beit. He had never been a strong man and, like Rhodes and Barnato, he stubbornly ignored the warnings of doctors who told him to rest. The trip to South Africa had been ill-advised; by pushing his failing powers to extremes he courted disaster. Rumours that he was not well filtered through from Rhodesia; they were dramatically confirmed when he returned to Johannesburg. On 8 January 1903, he suffered a slight paralytic stroke. The South African papers reported that he was dying: there was panic on the stock market.

Dr Jameson had left Johannesburg for the Cape but he returned immediately he heard of Beit's illness. Knowing that Jameson was on his way, Beit rallied. His meticulously ordered mind went to work. 'When he was told Dr Jameson was coming from East London', it was reported, 'he said: "Then he will be here in forty-three hours." Later, when informed that he was coming by special train Mr Beit gave the exact hour of his arrival in Johannesburg.'

Cared for by Jameson and three other doctors, Beit recovered sufficiently to return to England. Jameson travelled with him on the *Kildonan Castle* and saw him safely installed at 26 Park Lane before returning to South Africa. A few days later Beit left for Hamburg to be with his mother. He never fully recovered from the stroke.

According to Dr Hans Sauer, Beit's stroke was caused by the shock he received when he visited the Premier Diamond Mine near Pretoria. This newly discovered mine was not then under the control of De Beers and Sauer maintains that Beit saw its massive workings as a threat to his life's work. Others have repudiated this suggestion. There is reason to think that Beit's health was adversely affected by serious concern for his diamond interests, but his concern had nothing to do with the Premier Diamond Mine. The threat to his life's work came from a more bizarre source.

[6]

'Sir Starr Jameson', says G. A. L. Green, an editor of the *Cape Argus*, 'first let me into the secret that Rhodes's devoted friend and fellow millionaire, Alfred Beit, on the eve of his death . . . was firmly persuaded that artificial diamonds could be manufactured cheaply and that Kimberley and the diamond industry were faced with ruin. There appears to be little doubt that this "spectre of the night" aggravated Beit's failing health and hastened his end.' The authority Green quotes is sound. Dr Jameson was well acquainted with the state of Beit's health. He was also well versed in the extraordinary hoax which overshadowed the end of Beit's life.

There was nothing new about the possibility of artificial diamonds being manufactured. As early as 1880, the Kimberley papers had carried warnings of fake diamonds. Stories and articles on the production of such stones appeared at regular intervals. The peddling of artificial diamonds was regarded as a menace second only to the trade in stolen diamonds. The threat posed, however, was limited. Experienced diamond buyers could always recognise a synthetic stone; only the uninitiated were fooled. There was little danger of traffic in these stones undermining the diamond industry.

Not until the turn of the century did the production of artificial diamonds take a more serious turn. Then a highly reputable French physicist, M. Henri Moissan, announced that he had succeeded in manufacturing perfect tiny diamonds. Moissan was a Nobel prize-winner and his scientific colleagues had no hesitation in accepting his claim. All the same, the chance of Moissan's diamonds being marketed was remote. The cost of manufacturing them was so prohibitive that they could never rival the genuine article. A similar conclusion had been reached, some years earlier, when J. B. Hannay had produced microscopic diamonds from compressed charcoal.

Probably no more would have been heard of the French experiment had it not been for Henri Lemoine, an engineer employed by Moissan. In 1905 Lemoine let it be known that he had bettered his employer's achievement. He claimed that he had hit upon an entirely new formula for producing large, apparently flawless, artificial diamonds. Little was known of Lemoine, but so persuasive were his methods that he succeeded in convincing a number of British industrialists that his claim was genuine. As a result of the contacts he made in London, he was introduced to Alfred Beit's partner, Julius Wernher.

Wernher was no fool. Nor did he easily invite the attention of impostors. A hard-headed, conservative businessman, he kept himself so aloof that many people thought that Wernher was simply Beit's first name. Nevertheless Julius Wernher was completely taken in by Henri Lemoine. The Frenchman offered to produce a diamond for Wernher in his Paris workshop and a date was set for the demonstration. It was a huge success. From an electric furnace, installed in the basement of a Paris warehouse, Lemoine extracted a stone which, after it had been plunged into cold water, gave every appearance of being a diamond of great value. Wernher was sufficiently impressed to ask for a second demonstration and arranged for Alfred Beit and another diamond expert to attend. The same process was followed and another large diamond produced. Wernher and Beit were so convinced by the result that Beit agreed to sign a certificate which declared his satisfaction with the experiment.

Wernher then entered into negotiations with Lemoine. He agreed to finance a factory in the Pyrenees in which the Frenchman could continue his experiments. It is said that, by expending some £64,000 on this factory, Wernher hoped to ensure that Lemoine's discovery would not be exploited to the detriment of the diamond industry. If future experiments proved successful, De Beers would be in a position to purchase the Frenchman's secret formula. The amazing thing is that at no stage does Wernher appear to have sought scientific advice on Lemoine's methods. The only precaution he took was to have the so-called formula deposited in a London bank in a sealed envelope bearing his own and Lemoine's name. Only with their joint consent, or upon Lemoine's death, could this envelope be withdrawn.

Shortly after the factory had been established at Argelès, Wernher became alarmed by reports of Lemoine's activities there. He learned that the Frenchman had formed a company which was supplying local villages with electricity and that no use was being made of the furnaces installed for the diamond experiments. To make matters worse, Lemoine was insisting on further money advances and showed no enthusiasm for Wernher's proposal that new experiments be conducted before impartial witnesses. However, it was eventually agreed that a diamond expert from Kimberley should attend a demonstration at Lemoine's old laboratory in Paris.

Wernher contacted Dr Jameson—then Prime Minister of the Cape and a director of De Beers—who agreed that Lemoine's experiments should be thoroughly investigated. Jameson regarded this as a matter of the utmost importance. Not only the output of De Beers but the entire economy of the Cape would be jeopardised by the manufacture of artificial diamonds. In answer to Wernher's request he arranged for Francis Oats to pay a secret visit to Paris.

Oats was also a director of De Beers and, having started life as a Cornish miner, was a man with considerable mining experience.

On arriving in London, Oats was shown one of Lemoine's diamonds. The first thing that struck him was its uncanny resemblance to a natural Kimberley stone. He started for Paris not a little suspicious. Watching Lemoine's experiments in the dingy warehouse basement, his suspicions increased. The opportunities for a fraudulent substitution of real diamonds in place of those said to be manufactured seemed all too obvious. Finally, Oats performed some sleight of hand himself. Surreptitiously he slipped a genuine diamond into the furnace. According to Lemoine's theory the heat of the furnace should then have reached the point where any diamond would be reduced to a worthless piece of carbon. But Oats's diamond was retrieved unharmed. This was sufficient to convince Oats that the furnace was a fake and Lemoine was an impostor.

Wernher had no option but to commence legal proceedings. Lemoine was arrested and brought before a Paris magistrate. Unfortunately, the opposing counsels were unable to agree on the evidence to be submitted. Lemoine's lawyer wanted his client to conduct one of his experiments in public; Wernher insisted that the sealed formula in the London bank be opened and examined by experts. Argument on these points was protracted and eventually Lemoine was released on bail. He then disappeared.

In Lemoine's absence, Wernher obtained permission to open the sealed formula. The contents of the mysterious envelope produced much amusement. Nothing could be easier than the manufacture of diamonds, it seemed. According to Lemoine's delightfully simple instructions, a quantity of powdered carbon placed in a furnace, with the heat raised from 1,700 degrees to 1,800 degrees centigrade, would crystallise into a valuable gem stone. Parisians found this enchanting. 'They roar with laughter at Wernher's name,' reported Mrs Lionel Phillips. 'That is why the French won't buy Rand shares.'

Later Lemoine was rearrested. He was dining at home when the police came for him. He greeted them with marvellous aplomb. 'We are busy with our dessert, Messieurs', he said. 'Would you like to have something to eat yourselves or would you like a glass of Bordeaux?' He remained as civilised throughout his trial, but to little avail. Found guilty of fraud, he was sentenced to six years' imprisonment.

Beit never knew this. He had died two years earlier convinced that the days of De Beers Consolidated Mines were numbered.

[7]

Alfred Beit died at Tewin Water on 16 July 1906. He was fifty-three. Since his return from South Africa in 1903 he had never been really well. He continued with his work, he travelled to and from Germany, he worried about his own affairs and those he had taken over from Rhodes. The strain was more than his weak heart could take. It needed only Lemoine's artificial diamond to bring him to the point of collapse.

His last year had been particularly active and worrying. In 1905 trouble arose between France and Germany over Morocco. There was strong anti-German feeling in England and Beit fretted about this. Towards the end of the year he went to Paris where it was arranged for him to have an interview with the French Prime Minister, M. Rouvier. Encouraged by this meeting Beit went on to Potsdam where he was received by the Kaiser at the Neues Palais. Ostensibly this visit had been arranged to enable Beit to hand the German Emperor a copy of the catalogue of his art collection, but Beit was hoping to use the occasion to bring about a meeting between the Kaiser and M. Rouvier. He was not successful. The Kaiser appears to have confused him with Sir Ernest Cassel, King Edward VII's stockbroker friend, and asked him to use his influence to prevent England interfering in the quarrel between France and Germany. Although the Emperor attached a great deal of importance to this meeting, he undoubtedly had an exaggerated idea of Beit's political connections. Nevertheless, Edward VII was afterwards interested enough to send Lord Esher to get Beit's views on the interview. It was the only known occasion when Beit acted as a political go-between.

Before leaving on his European tour, Beit had been reminded of another political agent. In January 1905 he, as one of the Rhodes Trustees, had been served with a summons issued by Princess Catherine Radziwill. The Princess had served only sixteen months of her two year prison sentence. Ill-health and her trouble-making propensities had earned her an early release: it is said that her gaolers were pleased to get rid of her. No sooner had she arrived back in Europe than she had started proceedings against the Rhodes

Trustees. She claimed £1,400,000 on the grounds that Rhodes was
the father of a 'female child' delivered to her in December 1897 and
had employed her as his political representative. The summons
was issued after the Rhodes Trustees had refused to negotiate with
her. It was simply a bluff. There was no evidence to support her
claim and she was obliged to drop it before it came to court. If she
had indeed given birth to a child in 1897, she must quickly have
abandoned it. For the last twenty years of her life she lived in
America; when she died in New York in May 1941 the only daughters
she acknowledged were those from her marriage to Prince Wilhelm
Radziwill.

Absurd as the Princess's claims were, they must have worried
Beit. He was devoted to Rhodes's memory and the possibility of a
further scandal would have weighed heavily with him. The Princess's
assertions, added to the activities of Henri Lemoine, were hardly
conducive to the peace of mind which his doctors considered
necessary.

In April 1896 he went to Wiesbaden in Germany for treatment.
While there he was cared for by his mother, his brother Otto, and his
friend Seymour Fort. But there was no improvement in his condition. He returned to England at the beginning of July and died in his
brother's arms a week later.

He was buried in Tewin churchyard the following day. Born,
bred and buried a Christian, he remained essentially Jewish all his
life. His gentleness, his ready sympathy for the underdog and his
strong sense of family were derived as much from his race as from
his religion. He was sincerely mourned by men of all creeds. South
African celebrities flocked to the Tewin church to follow his coffin.
Conspicuous among the mourners was his mother—then in her
eighties—who had travelled from Germany to attend the funeral.
She sat on a hard wooden chair at the graveside. 'The crowd', says
Sir Lewis Michell, 'once thoughtlessly crowded out her view.
"Good friends", she said in a singularly pleasant voice, "stand
aside, if you please, I would see the last of my dear son."'

In life Beit had been guided by the example of Cecil Rhodes.
That example was also apparent when he died. Although not so
widely publicised, Alfred Beit's will was as high-minded as that of
Rhodes. Besides providing generously for his family, his public bene-
factions amounted to £2,000,000. He endowed universities in South
Africa. He left £1,200,000 for transport facilities, for scholarships

and for social services in Rhodesia. The Beit Trust has contributed greatly to the country named after Cecil Rhodes. It has provided the most imposing memorial to Alfred Beit—the Beit bridge, which spans the Limpopo and joins South Africa to Rhodesia.

There is, however, another memorial to Beit; more modest but, perhaps, more appropriate. A small bust of him on a low pedestal outside the Kimberley Public Library is inscribed: 'Write me as one that loved his fellow men.' Few self-made millionaires are worthy of such a tribute. Alfred Beit was.

CHAPTER TWENTY TWO

A FIGHTER TO THE END

JOSEPH BENJAMIN ROBINSON outlived them all. He had been almost middle-aged when Rhodes, Barnato and Beit arrived in South Africa as young men, yet they all died long before him. Twenty years after Beit's death, J. B. Robinson was still very much a South African personality; still unpopular, still capable of sparking off controversy, and still very, very rich.

But he was a disappointed man. This was the only thing—apart, of course, from money—that he had in common with the others: he died disillusioned, basically unhappy. Trite as it may be to say that money brings misery, it has to be said of the South African magnates. However the cliché is cloaked, it remains an inescapable fact. They were rich and their money brought them very little happiness. 'The men who dug for gold and diamonds and were lucky enough to find what they sought appeared to be unlucky in everything else,' says P. Tennyson Cole, who knew them all. 'So far as I could judge they were the unhappiest men on earth. A camarilla of spite, envy and hatred engulfed them fiercely on all sides immediately they grew rich.' J. B. Robinson, probably the richest of them, was also the least happy.

It was not jealousy on the part of others that caused his misery. His warped personality had much to do with it. If he was engulfed by envy, spite and hatred, it was his own envy, his own spite and his own hatred. All his life he was obsessed by his defeat as a diamond magnate. The huge fortune he made on the Rand never fully compensated for the humiliation he suffered in Kimberley. For years it had been his ambition to dominate the diamond industry; to establish himself as Rhodes's financial and political superior. His failure to do so left him embittered. For Rhodes he nursed an intense and lasting hatred.

Much of his later career was influenced, directly or indirectly, by his determination to even the score with Rhodes. While he considered it sound financial policy to ingratiate himself with President

Kruger, he also recognised that by siding with the Transvaal
Government he was directly opposing Rhodes. This made him
many enemies among the uitlanders. His hand was detected behind
much of the propaganda put out by the Kruger regime. For although, once he moved to London, he appeared to withdraw from
active participation in South African politics he continued to pull
strings. His influence was apparent in the two newspapers which he
financed in the Transvaal. Both these organs were extremely active
in the days before the Jameson Raid. 'His papers at Pretoria and
Johannesburg,' it was said, 'one of them openly subsidised as a
Government organ, were simply busy in their habitual work of
fostering disunion and discouragement in the Uitlander ranks.' To
bolster their effectiveness, Robinson then started a newspaper in
Cape Town and appointed one of Rhodes's vitriolic opponents as its
editor. Rhodes recognised the threat. On several occasions before
the Jameson Raid he expressed the fear that a revolution in the
Transvaal might exchange 'President Kruger for President J. B.
Robinson.'

Nobody greeted the news of Jameson's inglorious failure with
more enthusiasm than Robinson. His moment, it seemed, had at last
come. He seized it. In later years he liked to give the impression that
throughout the crisis created by the Raid he acted with statesmanlike detachment. He was in London at the time and the British
Colonial Secretary had consulted with him. The advice he gave was
—or so he says—calmly considered. Rhodes's supporters told another
tale. According to them he leapt at the opportunity to vilify Rhodes
in his newspapers. 'He himself', says Edmund Garrett, 'cabled from
Park Lane during the most disastrous period of the crisis hundreds
and thousands of pounds' worth of messages demonstrating, day
after day, that Mr Rhodes was fallen, must be fallen, ought to be
fallen, irretrievably and for ever.'

Morality was, of course, on his side but the personal nature of his
attacks made it clear that the downfall of Rhodes was his main
concern. Inspired as it was by hate, the campaign he waged had one
redeeming feature. This was his attitude towards Alfred Beit.
Robinson had no love for Beit. Not only did he resent his friendship
with Rhodes but he had never forgiven his former partner for buying
him out of the valuable Rand property they had shared. All the same,
he seems to have retained some respect for Beit's integrity. When the
editor of his Johannesburg newspaper reprinted a violent attack on

Beit, which Labouchere had published in *Truth*, Robinson was furious. 'It's all very well for a London newspaper, which does not know Beit, to impute these unworthy motives to him,' he told the editor. 'But you know he is not that sort of man. Besides, he was once my friend, and, though I disagree with his policy, I will not have a word that reflects upon his personality printed in any paper of mine.' Righteous indignation was not characteristic of Robinson: it says much for 'little Alfred's' reputation.

Nothing, however, could reconcile Robinson to Rhodes. Even after the Jameson Raid he continued to look for ways in which he could further his feud. More than anything he wanted to challenge the diamond monopoly held by De Beers Consolidated Mines. The slightest rumour of a new diamond discovery would send him or his agent—the unlikeable Dr Murphy—hurrying to the diggings. He made a number of unprofitable investments in worthless diamond mines and on one occasion came near to falling for a gigantic hoax.

In 1898 great excitement was caused in Kimberley by the report of a new diamond field which had been discovered at Rooidam, near Hopetown in the Cape. It was confidently predicted that this new discovery would prove so rich that the mines at Kimberley would become of secondary importance. The so-called discovery had been made by two prospectors and the first person they contacted was Cecil Rhodes. They offered to sell the property for £100,000. Rhodes was suspicious. He insisted that the workings be examined by experts from De Beers. The discoverers were reluctant to allow this. They claimed that if the experts' findings were unfavourable it would lessen their chances of offering the property elsewhere. Not surprisingly, this brought the negotiations with Rhodes to a close. The property was then offered to Robinson who, after having the mine hastily examined, handed over a large cheque for the Rooidam mine.

Eager to outwit Rhodes, Robinson had acted without his usual caution. The examination he had ordered had been far too superficial. Shortly after he had made his purchase, the Rooidam mine was exposed as a fake. The two 'discoverers' were charged with fraud and at a subsequent trial it was proved that they had employed a Hottentot labourer to feed a liberal supply of diamonds into the machine used for washing the ground from the mine.

However, Robinson had not been completely fooled. The exposure came before the hoaxers had time to cash his cheque and he quickly stopped payment. He lost nothing from the transaction: nothing except his hopes of defeating Cecil Rhodes.

[2]

In London Robinson continued to live in great style. Mean as he was, he spared no expense in the entertainments he 'lavished upon Society in Park Lane'. His dinners and concerts were elaborately staged and well attended, his guest list was often impressive. But despite his apparent generosity, his social amibitions were never fully realised. His hospitality was widely accepted but his circle remained strictly limited. There are few mentions of him in the social memoirs of the day; even the South African *nouveaux riches* tended to shun him.

If he was spoken of at all it was because of his money. His fortune became the subject of endless, often exaggerated, speculation. 'Money he has amassed beyond the dreams of avarice,' wrote W. T. Stead in the early 1900s. 'If he doubled his hoard it would not increase by one iota his ability to satisfy every wish, to secure every comfort, to gratify every caprice. If Mr Robinson had been born on the date printed in the margin of our Bibles as that of the creation of the world, and surviving all accidents of mortality, had lived down to our time, and drawn every twelve months since Adam a regular income of £1,000, he would not, even if he had never spent a penny, unless he had put his money out to usury, have accumulated the fortune with which he is popularly credited in the City and South Africa. What matters it to the master of so many millions the increment of wealth, the extent of which is already beyond his power to imagine or realise.' Robinson's riches were rated high indeed. It did nothing to increase his popularity.

Rightly or wrongly, his former friendship with President Kruger had earned him a bad name. He never lived it down. His support for a more enlightened Transvaal politician, however, brought its rewards. When Louis Botha became Prime Minister of the Transvaal in 1907 he was faced with a serious problem. The mine-owners of Johannesburg had imported Chinese labourers to work the gold mines and this, not surprisingly, had created bitter resentment both in South Africa and in England. Botha's first task upon assuming office was the repatriation of the Chinese mine-workers against the wishes of the mine-owners. Robinson was one of the few who

supported him by offering to help his government run the mines
with European and African labour. This broke the opposition to
Botha's repatriation policy and Robinson's assistance was duly
recognised. In June 1908 he was created a baronet.

As Sir Joseph Benjamin Robinson Bart, he had reason to congratulate himself. He had come a long way since the days when the
height of his ambition had been to be elected Mayor of Kimberley.
All the same, he was not allowed to forget his past. The Johannesburg
magnates were now more suspicious of him than ever; they took
great delight in spreading ill-natured stories about his early career.
Truthful or not, they were readily believed.

A little more substance was given to the gossip when, in 1911,
Lou Cohen published his scurrilous book, *Reminiscences of Kimberley*.
Based on a series of articles that had earlier appeared in the *Winning
Post*, the book dug up past scandals and contained some highly
libellous assertions. Little notice had been taken of the *Winning
Post* articles—which, if anything, were more outspoken—but the
book created something of a stir. Barney Barnato was the main
target for Cohen's attack but, in passing, he took some well-aimed
swipes at other Kimberley personalities. Many of those attacked
were dead by the time the book was published and those still living
were, for the most part, more amused than offended by Cohen's
snide comments. Dr Jameson is said to have roared with laughter at
the remarks about himself. However, Cohen made one serious
misjudgement. He singled out Sir Joseph Benjamin Robinson for
special attention. He should have known better. Much as he disliked
Robinson, he must have been fully aware of the 'Old Buccaneer's'
litigious tendencies.

The passing of the years had by no means sweetened Robinson.
He still dearly loved a legal tussle and was ready to go to court at the
drop of a hat, let alone a libel. In 1903, for instance, he had allowed
himself to be sued by a local tradesman in the County Court at
Dover simply so that he could prove a legal point. The tradesman
had supplied certain goods to Robinson's yacht *La Belle Sauvage*—
in which he achieved his greatest social success by entertaining
King Edward VII—and Robinson claimed that he could not be
held responsible for the payment of these goods. 'Mr Robinson', it
was reported, 'won the action, on a point of yachting law, that a
yacht owner is not responsible for debts incurred by his steward.
He has now, however, paid the tradesman his account, having fought

the action to prove the law on the matter.' With only this sort of exercise to amuse him, it was highly unlikely that Robinson would ignore Cohen's book. He was in South Africa when *Reminiscences of Kimberley* was published. As soon as a copy of the book was shown to him, he hurried back to London and sued Lou Cohen for libel.

For all its salacious innuendoes, there was much truth in Cohen's book. Had he been content to report what he knew from personal experience—and that was sensational enough—he might have been able to justify his stories. What he said about Robinson had been common gossip in Kimberley but, of course, gossip is not a matter for a court of law. One of the assertions contested by Robinson was that he had defrauded the Boers with whom he had traded in his early days in the Free State. This had been said, in a variety of ways, many times before but it was impossible to prove. Whatever the truth of the matter, the colourful version published by Cohen was undoubtedly false. Robinson knew that he stood a good chance of proving it false and spared no expense in preparing his case. Three of the most sought-after counsels in England—including the formidable Sir Edward Carson—were briefed to defeat Cohen.

The case was as good as won before it came to court. Cohen had neither the time nor the money to summon witnesses in his defence. Those he did call were, to say the least, highly suspect. Two of them made wild statements about Robinson's career on the river diggings and in Kimberley, but they failed to impress the court. Judgement was given in Robinson's favour and he was awarded £1,100 damages. The book was ordered to be withdrawn from circulation.

But that was not the end of the matter. Cohen's witnesses had claimed that Robinson had once been dragged through the river at Klip Drift for trading in diamonds with Africans and had been involved in I.D.B. in Kimberley. This Robinson was determined to disprove. He instituted criminal proceedings against Cohen for inciting the witnesses to perjury. It took him over two years to mount his case. He brought witnesses from South Africa—including old Stafford Parker, the former President of the Diggers Republic—to testify to his good character.

The case was heard at the Old Bailey at the beginning of 1914. Cohen had not a leg to stand on. It was shown that, not only had his witnesses lied, but that one of them, Emile Berger, had never set foot in South Africa. Sentencing Cohen, the judge said: 'I have no doubt that you, for your own purposes, wickedly suborned that man

Berger to tell deliberate lies, which he knew were lies. . . . I should pass on you a more severe sentence than was passed on him. But you are an old man, and I think I can meet the justice of the case by passing on you the same sentence as was passed on Berger. I cannot pass a less sentence than three years penal servitude.'

The reporting of this case involved Robinson in further litigation. Another digger threatened revelations about Robinson's past, but he was effectively silenced. Success at law meant a great deal to Robinson. He was no doubt delighted with his triumph over Cohen. But not always was he so clearly in the right. In the most important action he ever contested, judgement went against him. It cost him a small fortune and a coronet.

[3]

In 1915, at the age of seventy-five, Robinson decided that the time had come for him to retire. For years he had carried the burden of his mining interests practically single handed. His partnership with Maurice Marcus had long since been dissolved and, apart from Jan Langerman—who acted as his Managing Director in Johannesburg —he had no other close business associates. He was far too difficult to work with to make any lasting business relationships; even Langerman, nominally his South African deputy, was more of an employee than a partner. Robinson had always insisted upon manipulating the strings and making the decisions.

It pleased his ego to work alone; to be solely responsible for his undoubted success. But, gratifying as this was, it created problems. When the time came for him to retire, there was nobody to replace him. He had no option but to sell out lock, stock and barrel. He was forced to look for a buyer for his Witwatersrand companies. Finally he found one in Solly Joel.

Despite his widespread unpopularity, Robinson had retained a sort of friendship with the Barnato and Joel families. They lived close to each other in London and paid occasional social visits. This probably influenced Robinson's decision to sell to Solly Joel. Friendship, however, played little part in the bargaining which preceded the sale. Both Joel and Robinson were hard-headed businessmen, each determined to get the better of the other. The preliminary negotiations were stringent and protracted. Agreement was eventually reached in 1917 when Solly Joel paid Robinson roughly £4,500,000 for the Randfontein and Langlaagte gold mining companies. They were both convinced that they had made a good bargain. Robinson had.

It is said that when Solly Joel's engineers came to examine the Robinson properties they found them shockingly run down and badly mismanaged. An enormous sum was required to restore them to productivity and several years were to pass before they began to pay dividends. But Solly Joel was in for further shocks. His auditors soon discovered a serious discrepancy in the past expenditure of the companies. Examination of the records showed that Robinson had, to put it mildly, been less than honest in his dealings with his own shareholders.

The details of what became known as the 'Secret Profits Case' are complicated. What they amounted to, in essence, was that Robinson had been selling land and options to his companies at an enormous profit to himself. By anticipating the need for certain mineral interests, he had been able secretly to purchase parts of valuable Transvaal farms and resell them to his board of directors. On one occasion he had, for instance, secured a half share of the Waterval farm for £60,000 and, within a month, sold it to a company he controlled for £275,000, netting no less than £215,000 profit. From this profit £5,000 was deducted for expenses and commissions, leaving Robinson with a clear £210,000. 'No doubt the freehold was most valuable,' declared the Chief Justice of South Africa, 'and I assume that its acquisition, even at the price which Sir J. B. Robinson fixed, was beneficial to Randfontein Estates. But in the process of benefiting the company he was making £210,000 for himself, and that was a fact which it was necessary to conceal.' There were other facts which had been concealed, other large profits made in secret. Solly Joel had no hesitation in instituting proceedings.

The 'Secret Profits Case' opened in the South African Supreme Court in 1921. Fought by two vindictive millionaires, it became a *cause célebre*. Solly, demanding repayment of the unfair profits, briefed a capable team of lawyers to press his claim. Robinson put up a great show of outraged innocence. Although in his eighties, he still had plenty of fight left. Supported by an equally impressive array of legal advisers, Robinson spared no trick in his attempt to outwit Solly. Even his deafness he used to advantage. For it was noticed that while he appeared to have difficulty in hearing the questions put to him by the opposing counsel, he had no trouble in replying to his own lawyers. Not that this helped him much. Involved as the issues were, Robinson was plainly in the wrong. The judgement went against him.

As was to be expected, he went on fighting in the Appeal Courts; but without success. In the end he had to repay £462,000 and meet the legal costs. Altogether, the action and his subsequent appeals are said to have cost him £750,000.

Disastrous as was the outcome, the case could not have been heard at a more inopportune time. The following year it was announced that Lloyd George, the British Prime Minister, had seen fit to recommend Robinson for a peerage. This, coming so soon after a major scandal, was more than the House of Lords was prepared to take. During a lively debate in the Upper House, on 22 June 1922, the protests against Robinson's proposed elevation were both vehement and unprecedented. One after another their lordships rose in indignation to attack the would-be recruit to their ranks. Although the main criticism centred on the recent court case, a few older grievances were also raked up. Robinson's opposition to Rhodes and the uitlanders was by no means forgotten. The onslaught was led by Lord Harris, the Chairman of Rhodes's Gold Fields Company. He was ably supported by Lord Selborne, a former High Commissioner of South Africa, who accused Robinson of never having shown sympathy for England.

There was something ironic in the fact that, after so many years, it should have been the supporters of Cecil Rhodes who rubbed salt in the wounds inflicted by Barney Barnato's nephew.

At the end of the session, it was announced that the debate would be continued the following week. It was not. When their lordships met on 29 June, the Lord Chancellor read to them a letter which Robinson had sent to the Prime Minister:

'My dear Prime Minister,—I have read with surprise the discussion which took place yesterday in the House of Lords upon the proposed offer of a peerage to myself. I have not, as you know, in any way sought the suggested honour. It is now some 60 years since I commenced as a pioneer the task of building up the industries of South Africa. I am now an old man, to whom honours and dignities are no longer matters of much concern. I should be sorry if any honour conferred on me were the occasion for such ill-feeling as was manifested in the House of Lords yesterday, and, while deeply appreciating the honour which has been suggested, I would wish, if I may without discourtesy to yourself and without impropriety, to beg His Most Gracious Majesty's permission to decline the proposal.'

However, later he is said to have remarked: 'The title did not cost me a single penny.'

[4]

Having to decline the peerage was humiliating for Robinson. Newspaper cartoonists and gossip columnists had a great deal of fun at his expense. 'The Earl of Randfontein' became a music-hall joke. It was more even than the abuse-hardened Robinson could take. He had been on the point of buying himself a country house in England; now he abandoned this idea, closed his house in Park Lane, and retired to South Africa.

Before leaving London he created one further sensation by offering his valuable art collection—estimated to be worth £3,000,000 —for sale. The paintings had been in storage for some years and were now unpacked and hung in Christie's sale rooms. Illustrated catalogues were sent to leading art dealers and the sale was extensively advertised in Europe and America. It was said to be 'one of the greatest picture auction-sales of modern times'. A few days before the sale Robinson was pushed round the auction rooms in a wheel chair. He said later that when he saw his pictures hung against a background of rich red silk, he had doubts about parting with them. This was his excuse for placing such a high reserve price on the paintings. All the same, he considered his prices to be justified. 'The actual price I originally paid for the pictures has nothing to do with their value today,' he declared. 'The fact that a thing was bought cheaply does not necessarily mean that it should be sold cheaply. It would be impossible for me to take the pictures to South Africa, for example, but now I intend to turn them to good advantage.'

The sale has been described as 'unique in the annals of Christie's'. For, despite the fact that the collection included works by Constable, Gainsborough, Lawrence, Rembrandt, Hals and 'the cream of Millais's works', only a handful were sold. The auction rooms were packed with celebrities—from the Duke of Marlborough to John

McCormack, the tenor—but the prices set by Robinson put the paintings beyond the reach of the most ardent collectors or dealers.

Robinson returned to South Africa in high spirits. 'My grandmother lived to 103,' he declared, 'my two grandfathers lived to 97 and 93, and my brother died at 95, not long ago, so I still have a chance. . . . I am 83, and in spite of being a sufferer from rheumatism, I feel as game as a bantam.' And, denied his peerage, he could at least draw satisfaction from the fact that his daughter, Ida, had married the Italian Consul-General for South Africa and was now Countess (she later became Princess) Labia.

But the end was nearer than he thought. He lived out the last few years of his life in his rambling, heavily shuttered and closely guarded mansion, Hawthorndene, at Wynberg, near Cape Town. A crotchety, misanthropic old man, he rarely went out and received few guests. Those who visited the house during these latter days were never fully at ease; there was something indefinably mysterious about the place; it seemed to contain secrets, locked up in its many unused rooms. On rare occasions, when a party was given for his children, Robinson would receive the guests hunched up in a chair, dressed in a tussore-silk suit and wearing that symbol of his early Kimberley days, a white pith helmet. He never stayed long and would disappear before the party started.

He was practically stone deaf and stories of his meanness abounded. One often repeated tale has it that a friend, sitting next to him at dinner, plucked up courage to ask him for five pounds. Robinson appeared to listen to the request and then explained that he could hear nothing with that ear. Nothing daunted, the man got up, went to Robinson's other side, and asked for ten pounds. 'What?' exclaimed the millionaire. 'Ten pounds? You said a fiver just now.' The story has a familiar ring about it, but it could have originated with J. B. Robinson.

Shortly before he died he arranged for his biography to be published. Entitled *Memories, Mines and Millions*, it was supposedly written by Leo Weinthal, who had edited Robinson's pro-Kruger newspaper in Pretoria. It is very difficult to recognise the pugnacious, controversial Robinson in the statesmanlike figure who emerges 'half hero half angel' from its pages. Those who knew something of Robinson's early career were in no doubt about its veracity or its authorship. 'Weinthal', declared one, 'has timidly edited and issued the life of J.B.R., which is nothing more than a mass of self-advertisement and misrepresentation; the whole thing prepared and paid for by Robinson himself.' This eulogistic concoction was largely responsible for the last, most devastating, attack launched against the 'Old Buccaneer'.

He died, aged eighty-nine, on the morning of 30 October 1929. His final illness had lasted five weeks. 'Sir Joseph', said his doctor, 'made a most gallant fight right up to the end.' He was buried in the grounds of his Cape mansion. The local newspapers carried polite, conventional obituaries, based largely on his recently published biography.

A week later his will was published. Those who remembered the public benefactions of Rhodes and Beit, were appalled by its shabby sentiments. By no act did Robinson stand more condemned than by the publication of his last wishes. At once there was a change in the tone of the press. In one of the most vitriolic editorials ever published in South Africa, the *Cape Times* damned his memory.

Under the heading *Nil Nisi Malum*, it declared: 'He is in his grave, and the voice of his contemporaries is perforce silent about the evil which his long and unredeemed career compelled them, without known exception, to think of him. Thus debarred from speaking their minds about this dead millionaire (whose money is reputed credibly to run into the vast sum of some ten to twelve millions sterling) his contemporaries have at least been able to feel that they are under no compulsion to attempt the forlornly charitable task of saying anything good about him. . . . [The terms of the will] would be revolting enough if the sum disposed of was insignificant, but seem—perhaps without logical justification—much more scandalously repugnant when they are the deliberate intention of a man to whom the chances of his mortal life had brought a huge fortune. . . . It stinks, too, against public decency. This man owed the whole of his immense fortune to the chances of life in South Africa. He has not left a penny out of all his millions to any public purpose in the country which showered these immense gifts upon him. . . . His immunity against any impulse of generosity, private or public, was so notorious that the name of J. B. Robinson became during his lifetime proverbial for stinginess, not only in South Africa but wherever men of the world congregate together. . . . The evil which the dead man thus speaks of himself is terrible to contemplate. It will live in the records of South Africa for all time; and those who in the future may acquire great wealth in this country will shudder lest their memories should come within possible risk of rivalling the loathsomeness of the thing that is the memory of Sir Joseph Robinson.'

This was the answer of those who had been nauseated by the smugness and deceit of his biography.

His family tried to defend him. In a spirited reply—the first of many—his daughter, Countess Labia, claimed that his fortune had been 'grotesquely exaggerated'. She said that he had given money to private charities and paid heavily in income tax; so heavily that he felt he had discharged his duty to his country. But she did not say what his fortune was, nor is the exact amount known. It was a good deal less than the *Cape Times* estimate: but J. B. Robinson undoubtedly died a multi-millionaire.

'There was no man who led such a quiet and unassuming life with his family as my dear father,' his daughter concluded. 'He thought continually of the comfort of those around him, even down to the native boys in his service. At the last moment on his death-bed he said to those attending him, "You make me cry—you are all so good to me." I could say much more that would convince you that you have made a terrible and monstrous attack on the memory of one of the finest men who ever lived—even though his methods may have differed from those of the average man, as his experiences throughout a very long life certainly did.'

One can sympathise with her sentiments. Her letter was worthy of her father and on one point, at least, its truth is undeniable. The methods and experiences of J. B. Robinson did indeed differ from those of the average man.

EPITAPH

THE diamond magnates are all buried far from Kimberley—the town which saw their spectacular rise from obscurity to international fame. Yet their burial places are strangely appropriate to the personalities that Kimberley once knew.

Barney Barnato, whose chief concern was the advancement of his family, is buried in their midst, in the Jewish cemetery at Willesden in London.

Cecil John Rhodes, the visionary who gave his life over to the implementation of his grandiose ideals, is buried on 'The View of the World'—a starkly dramatic hilltop that overlooks the country that bears his name.

The modest, self-effacing Alfred Beit rests in a quiet English churchyard.

J. B. Robinson, who lived always for himself, lies in an inaccessible vault, protected by the tall iron railings that surround his home at Wynberg in the Cape.

In death, no less than in life, the diamond magnates retain their strongly divergent identities.

ACKNOWLEDGEMENTS

MOST books about early Kimberley are based on the memoirs of old diggers. I have found these books, and their sources, extremely unreliable. Biographers of the diamond magnates have relied almost entirely on such material and have often been led sadly astray. Important events in the lives of Cecil Rhodes, Barney Barnato and the Joel brothers have never been disclosed; there are no informed critical studies of J. B. Robinson and Alfred Beit.

In an attempt to explore this period more fully, I have had to return to contemporary sources. This has entailed, among other things, a detailed examination of the newspapers of the period. While fully aware of the pitfalls involved in working from press reports, I have found this material exceptionally informative. For most of the period dealt with in this book, Kimberley was served by three rival newspapers. These papers catered for a small, closely-knit community and any misreporting was quickly seized upon and corrected. Besides the coverage they gave to diamond company meetings, Government inquiries and law cases, their editorials were remarkably outspoken and all controversial issues were fully thrashed out. The correspondence they published was often as uninhibited and revealing as any private exchange of letters.

Newspaper research was, of course, supplemented by an extensive examination of unpublished documents and I am most grateful for the assistance I received in tracing such material. My chief debt is undoubtedly to Mr Theo Aronson, whose patient encouragement and expert advice was, in no small measure, responsible for the completion of this book. Anyone writing about early Kimberley soon calls on the assistance of Mrs Judy Hoare—a former Librarian of the town—and I, like others, have found her detailed knowledge of the Diamond City invaluable to my researches. In writing my brief sketch of the early diamond rush, I was fortunate in obtaining the enthusiastic co-operation of Mrs Marian Robertson. Few people

can know as much about this period as Mrs Robertson and she was
most generous in checking my account against the mass of material
she has collected for her forthcoming book on the pioneering days.
I am grateful, also, to Mr Percy Baneshik of Johannesburg who
kindly read my account of the shooting of Woolf Joel and verified it
from material he is collecting for a biography of Baron von Veltheim.
To Mrs Ann Marx of Johannesburg I owe my sincere thanks for the
loan of newspaper cuttings and other material kept by her great-
grandfather, Leopold Loewenthal—one of the colourful personalities of early Kimberley and Johannesburg. And I am greatly
indebted to Mrs N. MacFarlane (*née* Mackintosh) of Durban for
providing me with her eye-witness account of Barney Barnato's
drowning.

For his interest and advice, I must thank Mr A. P. Cartwright, the
historian of the South African gold fields, and for the loan of ur-
gently required books I am greatly indebted to Miss Joy Collier,
Miss Norah Henshilwood and Mr Peter Pitchford. While I can no
longer thank, I must acknowledge the kind interest and guidance
of the late Mr Basil Humphries of De Beers Museum in Kimberley.

I have received help from a number of public institutions and my
particular thanks are due to Mr P. J. van der Walt, of the Kimberley
Public Library, and his wife, Mrs E. van der Walt, who went to endless trouble in supplying me with a fascinating collection of photographs of early Kimberley. Valuable help in picture research was
also given by Mrs P. Stevens, of the Special Collections Department of Cape Town University, and by the staff of the South
African Library, Cape Town. I am grateful for the help I received
from Miss Joan Davies and the staff of the Cape Archives; Mr J. C.
Quinton and the staff of the Parliamentary Library, Cape Town;
Miss Anna Smith and the staff of the Africana Museum, Johannesburg; Dr Richard Liversidge and the staff of the McGregor Museum, Kimberley; and the staff of the British Museum and Colindale
Library in England.

I am extremely grateful to Mrs Mackie-Niven of Amanzi,
Uitenhage, for sending me extracts, and allowing me to quote, from
the correspondence of her father, Sir Percy Fitzpatrick. For permission to quote from other unpublished and published material
my thanks are due to: Dr A. M. Lewin Robinson and the Council of
the South African Library—the Merriman and Bower papers; Dr

Richard Liversidge and the McGregor Museum—the Stow papers; and the Van Riebeeck Society of Cape Town—*Selections from the Correspondence of J. X. Merriman 1870–1890*, edited by Phyllis Lewsen.

REFERENCES

UNPUBLISHED SOURCES

Merriman Letters. The Correspondence of J. X. Merriman. South African Library, Cape Town.

Stow MSS. Memoir of the formation of the De Beers Mining Company Limited, and its subsequent transformation into De Beers Consolidated Mines Limited with five Life Governors, by F. Philipson-Stow. Macgregor Museum, Kimberley.

Beet MSS. The Diamond Fields Pioneer Association: History of the Diamond Fields. Edited by George Beet. Kimberley Public Library.

Michell papers. The Memoirs, Diaries etc. of Sir Lewis Michell. Cape Archives Acc 540/29. Cape Town.

Fitzpatrick papers. The private papers of Sir Percy Fitzpatrick. In the possession of Mrs C. Mackie-Niven, Uitenhage, Cape.

Bower MSS. Reminiscences of Sixteen Years Service in South Africa, by Sir Graham Bower. South African Library, Cape Town.

NEWSPAPER ABBREVIATIONS

Ind. The Independent, Kimberley; *D.N. Diamond News,* Kimberley; *D.F.A. Diamond Fields Advertiser,* Kimberley; *D.T. Diamond Times,* Kimberley; *C.A. Cape Argus,* Cape Town; *C.T. Cape Times,* Cape Town; *E.P.H. Eastern Province Herald,* Port Elizabeth; *S.A. South Africa,* London; *W.P. Winning Post,* London.

PROLOGUE

p. 3 'where that came from . . .' *D.F.A.,* 5 Dec. 1885; 'Her Majesty possesses . . .' Boyle: *Cape For Diamonds,* p. 218.
p. 4 'During the time . . .' *Geological Magazine,* Jan. 1869.
p. 6 'Butchers, bakers, sailors . . .' Quoted by Rosenthal: *Gold.*
p. 7 'While the stuff . . .' Payton: *Diamond Diggings,* pp. 8–9; 'The quietude of these . . .' Quoted by Doughty: *Early Diamond Days,* p. 7.
p. 9 'In justice to Mr Parker . . .' Boyle: *Cape For Diamonds,* p. 58.
p. 10 'Through the straggling . . .' Ibid., p. 112.

CHAPTER ONE

p. 13 'nor is the name . . .' Emden: *Randlords,* p. 115.
p. 14 'Seated on the poop . . .' Pringle: *Narrative,* p. 7.
p. 16 'I know of course . . .' *Fitzpatrick papers.*

p. 17 'pretty stones . . .' and 'I gave a man . . .' *S.A.*, 23 Dec. 1893.

p. 18 'I brought up all . . .' Ibid.; 'I always found . . .' Quoted by Rosenthal: *Men's Millions*, p. 191.

p. 19 'We arranged a red . . .' Ibid.; 'It was very early . . .' *S.A.*, 23 Dec. 1893.

p. 20 'Diamond Merchant . . . having . . .' *D.N.*, 5 June 1872; 'the very picture . . .' and 'He was virtually . . .' Beet: *Grand Old Days*, pp. 147–8; 'Mr J. B. Robinson is known . . .' *D.N.*, 3 Oct. 1872.

p. 21 'I had a good deal . . .' *D.N.*, 8 May 1875.

p. 22 'The pliant whip . . .' *D.N.*, 3 Oct. 1872; 'having lately used . . .' *D.N.*, 10 Oct. 1872.

p. 23 'The secret of . . .' and 'In the course . . .' *D.N.*, 3 Oct. 1872; 'Mr J. B. Robinson . . . has been . . .' *C.A.*, 15 March 1873; 'Signor Barnato, The Great Wizard . . .' *D.N.*, 31 Oct. 1872.

CHAPTER TWO

p. 26 'What did you do . . .' Cohen: *Rem. of Kimberley.*

p. 28 'Signor Barnato, the Greatest . . .' *C.A.*, 14 Sept. 1872; 'The different tricks . . .' *D.N.*, 22 Oct. 1872; 'The cry is "Still . . ." ' *D.N.*, 21 Nov. 1872.

p. 29 'he appeared to advantage . . .' *W.P.*, 15 Feb. 1908; 'Signor Barnato's excellent . . .' *D.N.*, 28 Nov. 1872; 'If anybody wants . . .' and 'Fully appreciating . . .' *D.N.*, 26 Nov. 1872.

p. 31 'Too late, boy . . .' There are many variations of this story. See for example: Jackson: *The Great Barnato*, p. 22; Raymond: *Barnato*, p. 16; Emden: *Randlords*, p. 127; 'one of the jolliest . . .' Raymond: *Barnato*, p. 16.

CHAPTER THREE

p. 33 'Why did I come . . .' Maurois: *Cecil Rhodes*, p. 26; 'My father frequently . . .' McDonald: *Rhodes*, pp. 7–8.

p. 34 'A grubby little . . .' McDonald: *Rhodes*, p. 8; 'long-headed . . .' Ibid.; 'A typical schoolboy . . .' Michell: *Life*, vol. 1, p. 19.

p. 35 'a tall lanky . . .' Williams, B.: *Cecil Rhodes*, p. 11.

p. 36 'Shouldn't do that . . .' Ibid., p. 40.

p. 37 'People out here . . .' Lockhart: *Rhodes*, p. 46; 'To hear Rolleston . . .' Colvin: *Jameson*, vol. 1, p. 40; 'Ah, they . . .' Michell: *Life*, vol. 1, p. 28.

p. 38 'Mr Rhodes of Natal . . .' Payton: *Diamond Diggings*, p. 215; 'Holding to one . . .' Boyle: *Cape For Diamonds*, p. 173.

p. 39 'a nasty habit . . .' and 'the greatest proportion . . .' Williams, B.: *Cecil Rhodes*, p. 29.

p. 40 'Diamonds have only . . .' Ibid., p. 29; 'in his shirt sleeves . . .' McDonald: *Rhodes*, p. 22.

CHAPTER FOUR

p. 42 'His Excellency was . . .' Murray: *S.A. Reminiscences*, p. 200.

p. 44 'Ruin, financial ruin . . .' *Diamond Field*, 28 Nov. 1874; 'He hated the nigger . . .' Kiewiet: *Imperial Factor*, p. 55; 'The objects aimed at . . .'

Macmillan: *Sir Henry Berkly*, p. 209; 'Diggers from America...' Froude: *Two Lectures*.

p. 45 'taking illegal oaths...' Wilmot: *Southey*, p. 278.

p. 46 'It was apparent...' Williams, A.: *Some Dreams*, p. 191.

p. 47 'The insanely injudicious...' Kiewiet: *Imperial Factor*, p. 55; 'Hitherto I have abstained...' *D.N.*, 8 May 1875.

p. 48 'Troops arrived yesterday...' *E.P.H.*, 9 July 1875; 'Our readers cannot...' *D.N.*, 8 July 1875.

p. 49 'I have waited...' *D.N.*, 12 Aug. 1875.

p. 50 'With the threat...' Ibid.

pp. 50-1 'the stoutest horsewhip...' and 'to do the thing...' *D.N.*, 14 Aug. 1875; 'Mr Robinson was just...' and 'indulging in that...' Ibid.; 'In Danger of...' *Mining Gazette*, 13 Aug. 1875.

CHAPTER FIVE

p. 53 'There was an end...' Kiewiet: *Imperial Factor*, p. 58.

p. 54 'Mr Cecil Rhodes' and 'was the last...' *D.N.*, 6 Jan. 1876; 'some months ago...' *D.N.*, 27 Jan. 1876.

p. 55 'The Attorney General...' *D.N.*, 8 Jan. 1876.

p. 56 'My character was...' *Beet MSS*.

p. 57 'I have never forgotten...' Michell: *Life*, vol. 1, p. 64.

p. 58 'By all means...' Williams, B.: *Cecil Rhodes*, p. 39; 'Signor Barnato will give...' *D.N.*, 1 June 1876; 'a large audience...' *D.N.*, 3 June 1876.

p. 60 'It rather interested...' *W.P.*, 22 Feb. 1908; 'When a spot...' Ibid.

p. 61 'If you can make...' Cohen: *Rem. of Kimberley*, p. 55.

p. 62 'was introduction enough...' Ibid., p. 83; 'Barney Barnato was...' *W.P.*, 1 Aug. 1908. 'When I heard...' *W.P.*, 18 July 1908; 'the cradle of...' Cohen: *Rem. of Kimberley*, p. 61.

p. 63 'look in that...' Wilson: *Down the Years*, p. 140.

p. 64 'The heroine's lamentations...' Cohen: op. cit., p. 265. 'When we saw...' *Ind.*, 11 Jan. 1876; ''Ow the dogs...' Cohen: op. cit., p. 335.

p. 65 'be it from...' *D.N.*, 16 Sept. 1876.

p. 66 'Signor Barnato anticipating...' *D.N.*, 16 Sept. 1876; 'his popular entertainment...' *D.N.*, 17 Oct. 1876.

CHAPTER SIX

p. 67 'All that is revolting...' Little: *South Africa*, p. 43.

p. 68 'But Mr Dunkelsbuher...' *D.N.*, 22 Feb. 1876.

p. 70 'I can assure Mr Siddall...' *D.N.*, 24 Feb. 1876; 'In 1872 I bought some...' *S.A.*, 23 Dec. 1893.

p. 71 'A man of many...' Cole: *Vanity*, p. 185.

p. 72 'The Devil himself...' *W.P.*, 12 Sept. 1908. 'Could bunting have...' *D.N.*, 6 July 1876.

p. 73 'When it became...' Cohen: op. cit., pp. 240-1.

p. 74 'I purchased the *Independent* . . .' *S.A.*, 23 Dec. 1893; 'business on the lines . . .' Lockhart: *Rhodes*, p. 81.

p. 75 'Neither of these . . .' *E.P.H.*, 19 Dec. 1876; 'The verdict . . .' *E.P.H.*, 9 Feb. 1877.

p. 76 'I am not the proprietor . . .' *D.N.*, 30 Oct. 1879 (which gives details of the Hartley bond).

p. 77 'by the curiously . . .' *D.N.*, 9 Oct. 1880; 'Although he looked . . .' *W.P.*, 11 April 1908; 'already half-deserted . . .' Trollope: *South Africa*, p. 154; 'The many friends . . .' *D.N.*, 6 Oct. 1877.

CHAPTER SEVEN

pp. 78–80 For Trollope's descriptions of Kimberley see Trollope: *South Africa*, pp. 152–85.

p. 80 'Now Mr Butler . . .' Williams, B.: *Cecil Rhodes*, p. 39.

p. 81 'The impression he left . . .' Wirgman: *Storm & Sunshine*; 'He certainly had . . .' Michell: *Life*, vol. 1, p. 80.

p. 83 'It often strikes . . .' Quoted by Colvin: *Jameson*, vol. 1, pp. 50–1; 'Some people have . . .' Ibid., p. 76.

p. 84 'He had his views . . .' and 'He is accredited . . .' Warren: *On Veldt*, pp. 227–307; 'they could not afford . . .' Williams, B.: op. cit., p. 46.

p. 85 'His friends once . . .' Michell: *Life*, vol. 1, p. 67.

p. 86 'Here was a chance . . .' Cohen: op. cit., p. 265.

p. 87 For Lanyon Theatre incident see *Ind.*, 18 June 1878, and Jackson: *Great Barnato*, p. 44.

p. 90 'the plainest man . . .' and 'Perhaps I am . . .' Cole: *Vanity*, p. 183; 'Fanny Bees, Spinster' Jackson: *Great Barnato*, p. 46.

p. 91 'for the mere sake . . .' *D.N.*, 26 Nov. 1878; 'It is well known . . .' *D.N.*, 16 Dec. 1879.

p. 92 'After a little while . . .' *D.N.*, 12 Dec. 1878; 'the excellent organisation . . .' Ibid.; 'An edifying spectacle . . .' *D.N.*, 5 April 1879.

p. 93 'members who had . . .' *Ind.*, 29 Jan. 1879.

CHAPTER EIGHT

pp. 94–5 For Massett's impressions of Kimberley see *C.T.*, 8 April 1880.

p. 96 'I am. Then you are . . .' *D.F.A.*, 30 July 1879.

p. 97 'This did not meet . . .' Ibid.; 'Stockdale Street . . .' *Ind.*, 29 July 1879.

p. 98 'I should imagine the income . . .' *C.T.*, 8 April 1880.

pp. 99–100 'Whether Mr Robinson . . .' *Ind.*, 15 Dec. 1879; 'Mr Birbeck was . . .' *Ind.*, 22 Dec. 1879; 'It is to be hoped . . .' *D.N.*, 27 Dec. 1879; 'Those who opposed . . .' *D.N.*, 28 Feb. 1880; 'Mr J. B. Robinson who . . .' *D.N.*, 7 Oct. 1880.

p. 101 For scenes in the council chamber see *D.N.* and *Ind.*, 7–15 Oct. 1879.

CHAPTER NINE

p. 103 'such claimholders as . . .' *Ind.*, 16 June 1877.

p. 104 'Mr Robinson found . . .' *Ind.*, 16 June 1877; 'gone to the expense . . .'

Trollope: *South Africa*, p. 165; 'Many stories of fortunes . . .' *C.T.*, 8 April 1880.

p. 105 'To secure the . . .' *Ind.*, 22 Oct. 1881.

p. 107 'Another Amalgamation . . .' *Ind.*, 29 May 1880; 'his brass . . .' *Ind.*, 4 June 1880.

p. 108 'Messrs Barnato . . .' *Ind.*, 10 March 1881.

p. 109 'My ancestors . . .' McDonald: *Rhodes*, pp. 1-2.

p. 110 'Mr Barnato spoke . . .' *Ind.*, 26 Feb. 1881.

p. 111 'Mr Robinson's supporters . . .' *Ind.*, 15 March 1881; 'I remember his . . .' Michell: *Life*, vol. 1, p. 91; 'I heard several . . .' *Ind.*, 10 May 1881.

p. 112 'Mr Robinson was received . . .' *Ind.*, 15 May 1881; 'I am no Negrophilist . . .' *Ind.*, 9 Feb. 1881.

p. 113 'Rhodes and the Kimberley . . .' Williams, B.: op. cit., p. 63; 'It is generally admitted . . .' *Ind.*, 10 May 1881.

p. 114 'received with cries . . .' *Ind.*, 27 July 1881; 'hated Rhodes . . .' Barlow: *Almost in Confidence*, p. 48.

CHAPTER TEN

p. 117 'everything by turns . . .' *D.F.A.*, 20 Feb. 1870; 'I found the place . . .' *W.P.*, 10 April 1909.

p. 118 'Even people doing . . .' Trollope: *South Africa*, p. 176; 'Why try . . .' Ibid.

p. 119 'The Barnato Company's . . .' *Ind.*, 16 June 1881; Barnato Company report and comment see *Ind.*, 7 July 1881.

p. 120 'The Company has been . . .' *Ind.*, 12 July 1881; 'There is every . . .' *Ind.*, 7 July 1881; 'to Mr H. Barnato . . .' Ibid.

p. 121 'As for the moneyed . . .' Little: *South Africa*, p. 43; 'Every species . . .' Dixie: *Land of Misfortune*, p. 286; 'had for its object . . . *Ind.*, 25 March 1880.

p. 122 'The licensed diamond . . .' Angove: *Early Days*, p. 180.

p. 123 'The Kaffirs were . . .' *W.P.*, 24 March 1906.

p. 124 'I doubt if any . . .' *W.P.*, 3 Oct. 1908; 'After the laying . . .' *Ind.*, 16 July 1881; 'Among the passengers . . .' *Ind.*, 13 Aug. 1881.

p. 125 'Few men have been . . .' *Ind.*, 15 Aug. 1881.

CHAPTER ELEVEN

p. 126 'secured the key . . .' *Stow MSS.*; 'I have never . . .' Millin: *Rhodes*, p. 66; 'When Mr Rhodes . . .' *Stow MSS.*

p. 127 'as an advance . . .' Williams, G.: *Diamond Mines*, p. 280.

p. 128 'I hope you won't . . .' *Beet MSS.*; 'The silent, selfcontained . . .' Cohen: op. cit., p. 20; 'They shared . . .' Colvin: *Jameson*, vol. 1, p. 79.

p. 129 'Open the enclosed . . .' Michell: *Life*, vol. 1, p. 137.

p. 130 'The Oxford system . . .' Williams, B.: op. cit., p. 42; 'I can see it . . .' *Stow MSS.*; 'De Beers mine is . . .' *Ind.*, 14 March 1882.

p. 131 'Owing to the . . .' *Ind.*, 10 April 1882.

p. 132 'Mr Rhodes's appointment . . .' *Ind.*, 19 Sept. 1881; 'The reckless way . . .' *D.F.A.*, 12 April 1882.

p. 133 'I had been intimidated . . .' Graumann: *Rand Riches*, pp. 3–4; 'As regards the system . . .' *D.F.A.*, 12 April 1882.

p. 134 'Objections have been . . .' Quoted by Weinthal: *Men, Mines*, p. 82.

p. 135 'Mr Robinson is a man . . .' *D.F.A.*, 21 April 1882; 'The 19th of April . . .' *D.F.A.*, 28 June 1882.

p. 136 'I decidedly objected . . .' Mathews: *Incwadi Yami*, p. 216; 'A law of exceptional . . .' Churchill: *Men, Mines and Animals*, pp. 45–6.

p. 137 'We had great difficulty . . .' *S.A.*, 23 Dec. 1893; 'in his baby . . .' *Ind.*, 4 Feb. 1884.

p. 138 'By their efforts . . .' *Ind.*, 27 Sept. 1883.

CHAPTER TWELVE

p. 140 'the gross irregularity . . .' *Ind.*, 7 Nov. 1882.

p. 141 'A mode of payment . . .' Ibid.; 'To consider removing . . .' *Ind.*, 24 Oct. 1882.

p. 142 'Mr H. Barnato came . . .' *Ind.*, 7 Nov. 1882; 'the tribe and supporters . . .' *D.F.A.*, 7 Nov. 1882.

p. 143 'Dr Murphy will . . .' *Ind.*, 8 Feb. 1884.

p. 144 For Murphy-Isaac Joel encounter see *Ind.*, 21, 23 & 26 Feb. 1884.

pp. 144–7 For Isaac Joel's arrest and trials see *Ind.* and *D.F.A.*, 24 March, 3, 21, 28 April and 8 May 1884.

pp. 145–50 For Barnato–Robinson–Fry affair see *C.A.*, 5–6 May 1885 and *Ind.*, *D.F.A.* and *D.T.*, 6–7 May 1885.

p. 151 'As he, in his gleeful . . .' Cumberland: *What I think*, p. 68. For attack on J. B. Joel see *Celebrities in Glass Houses*, *W.P.* pamphlet N.D.

p. 152 'Mr Joel underwent . . .' Felstead: *In Search*, pp. 127–8; 'Men sometimes say . . .' Quoted by Joel: *Ace of Diamonds*, p. 212.

CHAPTER THIRTEEN

p. 153 'All this to be . . .' Williams, B.: op. cit., p. 55; 'Stay with me . . .' and 'There are very . . .' Michell: *Life*, vol. 1, p. 135.

p. 154 'Politics to me . . .' *Merriman Letters*; 'You will see . . .' Lewsen: *Correspondence*, p. 167; 'Well! All I can . . .' Ibid.

p. 155 'I look upon this . . .' Williams, B.: op. cit., p. 73; 'Blood must flow . . .' Ibid., p. 82.

p. 156 'That young man . . .' Ibid., pp. 86–7; 'When I am in Kimberley . . .' Sauer: *Ex-Africa*, p. 119.

p. 157 'I have always been . . .' *D.F.A.*, 15 March 1886; 'Amalgamation steadily . . .' *Merriman Letters*.

p. 159 'Fitzpatrick, you knew him . . .' Fitzpatrick: *S.A. Memories*; 'I was one . . .' Interview with Frank Harris: *John Bull*, 28 July 1906.

p. 161 'I just did . . .' and 'When I reached . . .' and 'I got something . . .' Ibid.

p. 163 'Of all the men . . .' Phillips: *Reminiscences*, p. 87; 'At first he worked . . .' *Beet MSS.*

p. 164 ' "Hullo" said Rhodes . . .' Quoted by Fort: *Alfred Beit*, p. 72.

p. 165 'Ask little Alfred,' Lockhart: *Rhodes*, p. 75.

CHAPTER FOURTEEN

p. 166 'The streets of Kimberley . . .' *D.F.A.*, 30 Nov. 1885.
p. 168 'The decline in the . . .' Mathews: *Incwadi Yami*, p. 254; 'It is the fashion . . .' *Ind.*, 21 Feb. 1881.
p. 169 'Who were to search . . .' and 'the obnoxious . . .' Angove: *Early Days*, p. 183.
p. 170 'The ruinously low . . .' *Ind.*, 3 July 1883.
p. 171 'It was not only . . .' Sauer: *Ex-Africa*, p. 78.
p. 172 'The fact is that . . .' *D.T.*, 14 May 1885; 'You cannot drown . . .' Boyle: *Cape For Diamonds*, pp. 376–7.
p. 173 'During your absence . . .' *Merriman Letters*.
p. 174 'message of salvation' Lewsen: *Correspondence*, p. 202; 'Personally I am . . .' Ibid., p. 204.
p. 175 'They must be proud . . .' *D.T.*, 4 May 1885.
p. 176 'Mr Robinson is virtually . . .' *Merriman Letters*.
p. 177 'I do not see . . .' Ibid.; 'The Central people . . .' Ibid.
p. 178 'I learned in London . . .' *D.F.A.*, 8 April 1886; 'Moulle was furious . . .' Lewsen: *Correspondence*, p. 208.
p. 179 'and that he did not . . .' Ibid., p. 209; 'We felt that . . .' *D.F.A.*, 7 May 1887.

CHAPTER FIFTEEN

p. 182 'My principle object . . .' *D.T.*, 17 July 1884; 'I cannot allow this occasion . . .' Ibid.
p. 183 'One of the first acts . . .' *Ind.*, 7 Nov. 1882; 'any deficit . . .' *Ind.*, 22 Oct. 1884.
p. 184 'Fortunately for us . . .' *D.T.*, 17 July 1884; 'I have just returned . . .' *D.F.A.*, 16 Dec. 1885.
p. 185 'One day he came . . .' *John Bull*, 4 Aug. 1906.
p. 186 'When I left London . . .' *D.F.A.*, 8 July 1887.
pp. 187–8 'the Transvaal Navy . . .' Ibid.

CHAPTER SIXTEEN

p. 189 'Such an immense number . . .' Payton: *Diamond Diggings*, p. 26; 'The centre of a country . . .' Boyle: *Cape For Diamonds*, pp. 154–5.
p. 190 'I read a very good . . .' Vindex: *Speeches*.
p. 191 'You young thieves . . .' Williams, B.: op. cit., p. 99.
p. 193 'a cunning little Jew . . .' *W.P.*, 22 Feb. 1908; 'Well, Mr Rhodes, . . .' Williams, G.: *Diamond Mines*, p. 287; 'You may tell . . .' Ibid.; 'You know the story . . .' Vindex: *Speeches*, p. 750.
p. 195 'They had all . . .' *D.F.A.*, 22 Sept. 1887.
p. 196 'It might be good enough . . .' Lockhart: *Rhodes*, p. 115; 'You can go and offer . . .' Vindex: *Speeches*.
p. 197 'Mr Barnato is only . . .' *D.F.A.*, 4 Nov. 1887.
p. 198 'Just think of it . . .' Vindex: *Speeches*; 'It took Barnato's . . .' *Beet MSS*.
p. 199 'Rhodes only beat me . . .' Jackson: *Great Barnato*, p. 78; 'Barnato and myself . . .' *Beet MSS*.

p. 200 'If only we can . . .' Vindex: *Speeches*; 'I really do not . . .' Ibid.

p. 201 'Mr Beit did not . . .' *John Bull*, 28 July 1906; 'I'll tell you what . . .' Vindex: *Speeches*; 'Oh, that's all right . . .' Williams, B.: op. cit., pp. 101–2; 'You've staked . . .' Fitzpatrick: *S.A. Memories*, p. 34.

pp. 202–3 'Well, you've had your . . .' Williams, B.: op. cit., p. 102.

CHAPTER SEVENTEEN

p. 207 'Aren't those just . . .' Lockhart: *Rhodes*, p. 120; 'The day that man . . .' Chilvers: *De Beers*, pp. 52–68.

p. 208 'There was considerable . . .' *Stow MSS*; 'He said offers . . .' Ibid.; 'These are the facts . . .' Vindex: *Speeches*.

p. 209 'When you have been . . .' Williams: op. cit., p. 103; 'And tonight . . .' Fitzpatrick: *S.A. Memories*, p. 32; 'If they do not . . .' Vindex: *Speeches*.

p. 210 'Men are being . . .' Raymond: *Barnato*, p. 88.

p. 211 'Since the time . . .' and 'The powers of the . . .' Michell: *Life*, vol. 1, p. 185.

p. 212 'This town . . .' Wilson, S.: *S.A. Memories*, p. 14.

CHAPTER EIGHTEEN

p. 216 'A discovery has . . .' and 'Pooh, it is . . .' *S.A.*, 23 Dec. 1893.

p. 217 'Beit arranged to stay . . .' *Fitzpatrick papers*.

p. 218 'Beit told . . .' Ibid.; 'I took some stuff . . .' *S.A.*, 23 Dec. 1893.

p. 219 'This served me . . .' Ibid.

p. 221 'He made liberal . . .' *Fitzpatrick papers*.

p. 222 'Come along Abe . . .' Taylor: *Pioneer*, p. 110; 'I am on very . . .' *S.A.*, 23 Dec. 1893.

p. 223 'I remember a concert . . .' Glover: *Four Continents*, p. 192.

CHAPTER NINETEEN

p. 224 'I came for a visit . . .' Emden: *Randlords*, p. 136.

p. 225 'Beit don't like . . .' Cohen: *Rem. of Johannesburg*, p. 58; 'What a devil of a . . .' Lewsen: *Correspondence*, p. 283.

p. 226 'You see I beat . . .' Cohen: *Johannesburg*, p. 58; 'He had no conception . . .' *South African Mining Journal and Financial News*, 19 June 1897.

p. 227 'I take no risks . . .' Emden: *Randlords*, p. 299.

p. 228 'Everybody who passes . . .' Raymond: *Portraits*, p. 31; 'I did get . . .' Cohen: *Johannesburg*, p. 269; 'I'm beautiful, bountiful . . .' Quoted by Jackson: *Great Barnato*, p. 184.

p. 229 'The Mansion House . . .' *The Times*, 8 Nov. 1895.

p. 230 'an act of suicidal . . .' Emden: *Randlords*, p. 143.

p. 231 'Mr Barnato, you are . . .' Emden: *Randlords*, p. 143; 'all our landed properties . . .' Colvin: *Jameson*, vol. 2, p. 148; 'No one else . . .' Raymond: *Barnato*, p. 191.

p. 232 'I can't forget . . .' Ibid.

p. 233 'They're after me . . .' Jackson: *Great Barnato*, p. 230.

pp. 233–8 For accounts of Barney Barnato's death see Jackson: *Great Barnato*;

Raymond: *Barnato*; Lewinsohn: *Barney Barnato*; Joel: *Ace of Diamonds*.

p. 234   'He was sitting . . .' Private information from Mrs N. MacFarlane (Nellie Mackintosh) of Durban.

p. 235   'We were told . . .' Ibid.; 'Always wind up . . .' Cumberland: *What I think*, p. 81.

p. 236   'some day that . . .' Raymond: *Barnato*, p. 168.

p. 238   'I suppose you thought . . .' Le Seuer: *Cecil Rhodes*, p. 19; 'to a better world . . .' Joel: *Ace of Diamonds*, quoted p. 71.

p. 241   'squared for a fiver' Jackson: *Great Barnato*, p. 238; 'If that is . . .' Joel: *Ace of Diamonds*, quoted p. 76.

p. 243   'I must say I am . . .' Emden: *Randlords*, p. 149.

p. 244   'I got you off . . .' Ibid., p. 149.

p. 246   'With the exception . . .' *W.P.*, 24 March 1906.

p. 248   'Never let a man . . .' Cumberland: op. cit.

CHAPTER TWENTY

p. 249   'It may be relied . . .' *D.F.A.*, 17 July 1886.

p. 250   'certainly one of the . . .' *D.F.A.*, 10 May 1884; 'some of the thorns . . .' *D.F.A.*, 28 June 1884.

p. 251   'with lots of specimens . . .' *D.F.A.*, 30 July 1886; 'He listened to what . . .' Sauer: *Ex-Africa*, p. 111.

p. 252   'You see, everyone . . .' Ibid., p. 112; 'Needless to remark . . .' *D.F.A.*, 14 Aug. 1886.

p. 253   'I'm off' Colvin: *Jameson*, p. 81; 'Buy a seat . . .' Fitzpatrick: *S.A. Memories*, p. 88; 'careless of anything . . .' Colvin: op. cit.

p. 254   'You have been . . .' Ibid.; 'Ah, Barney, . . .' Ibid.

p. 256   'When at last . . .' Williams, B.: op. cit., p. 150.

p. 257   'Has anyone else . . .' Plomer: *Cecil Rhodes*, p. 64.

p. 258   'If he cannot say . . .' Michell: *Life*, vol. 2, p. 119.

p. 259   'I hate Kruger . . .' *Bower MSS.*; 'Old Jameson . . .' Williams: op. cit., p. 271; 'In times of . . .' *D.F.A.*, 13 Jan. 1896.

p. 260   'It will take . . .' Wilson, S.: *S.A. Memories*, p. 30; 'I am going . . .' Jourdan: *Cecil Rhodes*, p. 83.

p. 261   'The problem of . . .' Baker: *Cecil Rhodes*, p. 22; 'The big and simple . . .' Ibid.

p. 262   'This is a . . .' Cumberland: op. cit., pp. 27-8; 'He seemed to have . . .' Jourdan: *Cecil Rhodes*, p. 20.

p. 263   'We were all . . .' Le Seuer: *Cecil Rhodes*, p. 190; 'Rhodes raved and stormed . . .' Johnson: *Great Days*, p. 105; 'He liked a man . . .' Jourdan: op. cit.

p. 265   'She glided into . . .' Jourdan: Ibid., pp. 86-7.

p. 266   'I paid her bills . . .' *C.A. Weekly*, 12 Feb. 1902.

p. 267   'clearly an absolute . . .' Ibid.; 'Damn that woman! . . .' Le Seuer: *Cecil Rhodes*, p. 312.

p. 268   'He gave his . . .' *The Cape*, 5 Feb. 1909.

p. 269   'So it came to pass . . .' Stead: *Will and Testament*; 'Everything in the world . . .' and 'From the cradle . . .' Millin: *Rhodes*, p. 155.

p. 270 'His face was bloated . . .' Le Seuer: *Cecil Rhodes*, p. 305; 'It is a marvel . . .' Ibid., p. 309.

p. 271 'Turn me over . . .' Millin: *Rhodes*, p. 351; For Rhodes' will see Michell: *Life*, vol. 2, Appendix.

CHAPTER TWENTY-ONE

p. 273 'All that Beit . . .' Fort: *Alfred Beit*, p. 58.

p. 274 'Beit's was the . . .' Quoted by Fort: op. cit., pp. 92–3.

p. 276 'I thought of only . . .' and 'For the last . . .' Quoted by Cartwright: *The Corner House*, p. 123; 'Rhodes could never . . .' Phillips, L.: *Reminiscences*, p. 88.

p. 277 'What's the use . . .' Leonard: *How we made Rhodesia*.

p. 278 'a considerable sum . . .' Fitzpatrick: *S.A. Memories*, p. 77; 'Yes, it is . . .' Ibid.; 'He does not disguise . . .' *S.A.*, 19 March 1892.

p. 279 'quite crushed . . .' Wilson: *S.A. Memories*, p. 66; 'I never saw anyone . . .' Phillips, F.: *Recollections*, p. 79; 'Mr Beit and several . . .' Quoted by Beit: *Will and Way*, p. 22.

p. 280 'ogre and business . . .' Emden: *Randlords*, p. 152; 'Mr Beit played a prominent . . .' Quoted by Beit: op. cit., p. 20; 'a cross between . . .' Fort: op. cit., p. 155.

p. 281 'We were seated . . .' *John Bull*, 28 July 1906; 'Perhaps the most . . .' *Illustrated London News*, 21 July 1906; 'He had a real . . .' *John Bull*, 28 July 1906.

p. 282 'Beit's artistic tastes . . .' *Michell MSS.*; 'He enjoyed the society . . .' Fort: op. cit., p. 158; 'Mr Beit? Beit . . .' *C.A.*, June 1969.

p. 283 'I called out . . .' Taylor: *Pioneer*, p. 75; 'his profound . . .' Fort: op. cit., p. 88; 'Beit, although a German . . .' Phillips, L.: *Reminiscences*, p. 141.

p. 284 'In this emergency . . .' *Michell papers*; 'Yes, but don't . . .' Quoted by Rosenthal: *Men's Millions*, p. 74; 'He always takes . . .' *S.A.*, 7 Feb. 1903.

p. 285 'Beit replied simply . . .' *Michell papers*; 'faithful lieutenants . . .' Wilson: *S.A. Memories*, pp. 283–4.

p. 286 'When he was told . . .' *S.A.*, 17 Jan. 1903; 'Sir Starr Jameson . . .' Green: *An Editor*, p. 122.

p. 289 'They roar . . .' Quoted by Gutsche: *No Ordinary Woman*, p. 224; 'We are busy . . .' Green: op. cit., p. 122.

pp. 290–1 For Princess Radziwill's attempt to sue the Rhodes Trustees see *Michell papers*; 'The crowd once . . .' Ibid.

CHAPTER TWENTY-TWO

p. 293 'The men who dug . . .' Cole: *Vanity Varnished*, p. 175.

p. 294 'His papers at Pretoria . . .' Garrett: *African Crisis*, p. 127; 'President Kruger for . . .' Lockhart: *Rhodes*, p. 308; 'He himself cabled . . .' Garrett: op. cit., p. 127.

p. 295 'It's all very well . . .' *Pall Mall Gazette*, 8 Feb. 1903.

p. 296 'lavished upon Society . . .' Garrett: op. cit., p. 126; 'Money he has amassed . . .' Quoted by Rosenthal: *Men's Millions*, p. 70.

p. 297 'Mr Robinson won . . .' *S.A.*, 17 Jan. 1903.

p. 298 'I have no doubt . . .' *S.A.*, 28 Feb. 1914.

p. 300 'No doubt the freehold . . .' Quoted by Joel: *Ace of Diamonds*, pp. 126–7.

p. 301 'My dear Prime Minister . . .' *S.A.*, 7 July 1922.

p. 302 'the title did not . . .' Emden: *Randlords*, p. 121; 'The actual price . . .' *S.A.*, 13 July 1923.

p. 303 'My grandmother . . .' Ibid.; 'What? Ten pounds? . . .' Emden: op. cit., p. 122; 'Weinthal has timidly . . .' *Fitzpatrick papers*.

p. 304 'Sir Joseph made a . . .' *C.T.*, 31 Oct. 1929; 'He is in his grave . . .' *C.T.*, 7 Nov. 1929.

p. 305 'There was no man . . .' *S.A.*, 15 Nov. 1929.

BIBLIOGRAPHY

Angove, John. *In The Early Days*. Handel House Ltd., Kimberley & Johannes-
burg, 1910.
Baker, H. *Cecil Rhodes by his Architect*. Oxford University Press, 1934.
Barlow, A. G. *Almost in Confidence*. Juta & Co., Cape Town, 1952.
Beet, George. *Grand Old Days of the Diamond Fields*. Maskew Miller, Cape
Town, 1931.
Beit, Sir Alfred. *The Will and the Way*. Longmans, London, 1957.
Boyle, F. *To The Cape For Diamonds*. Chapman & Hall, London, 1873.
Cartwright, A. P. *The Corner House*. Purnell, Cape Town, 1965.
Chilvers, H. A. *The Story of De Beers*. Cassell, London, 1939.
Churchill, Lord Randolph. *Men, Mines and Animals in South Africa*. Sampson
Low, London, 1892.
Cohen, Louis. *Reminiscences of Kimberley*. Bennett & Co., London, 1911.
— *Reminiscences of Johannesburg & London*. Robert Holden, London, 1924.
Cole, P. Tennyson. *Vanity Varnished*. Hutchinson, London, 1931.
Colvin, Ian. *Life of Jameson*. Edward Arnold, London, 1922.
Cumberland, Stuart. *What I Think of South Africa*. Chapman & Hall, London,
1896.
Dixie, Lady Florence. *In The Land of Misfortune*. Bentley, London, 1882.
Doughty, Oswald. *Early Diamond Days*. Longmans, London, 1963.
Emden, Paul H. *Randlords*. Hodder & Stoughton, London, 1935.
Felstead, S. T. *In Search of Sensation*. Robert Hale, London, 1945.
Fitzpatrick, Sir J. Percy. *South African Memories*. Cassell, London, 1932.
Fort, G. S. *Alfred Beit*. London, 1932.
Froude, J. A. *Two Lectures on South Africa*. Longmans, London, 1880.
Garrett, Edmund. *The Story of an African Crisis*. A. Constable, London,
1897.
Glover, Lady E. *Memories of Four Continents*. Seeley, Service & Co., London,
1923.
Graumann, Sir Harry. *Rand, Riches and South Africa*. Simpkin Marshall,
London, 1936.
Green, G. A. L. *An Editor Looks Back*. Juta, Cape Town, 1947.
Gutsche, Thelma. *No Ordinary Woman*. Timmins, Cape Town, 1966.
Jackson, Stanley. *The Great Barnato*. Heinemann, London, 1970.
Joel, Stanhope. *Ace of Diamonds*. Muller, London, 1958.
Johnson, Frank. *Great Days*. G. Bell, London, 1940.
Jourdan, Philip. *Cecil Rhodes*. Bodley Head, London, 1910.

Kiewiet, C. W. de. *The Imperial Factor in South Africa*. Cambridge University Press, 1937.

Le Seuer, Gordon. *Cecil Rhodes*. John Murray, London, 1913.

Lewinsohn, Richard. *Barney Barnato*. Routledge, London, 1937.

Lewsen, Phyllis. *The Correspondence of J. X. Merriman 1870–1890*. Van Riebeeck Society, Cape Town, 1960.

Little, J. S. *South Africa: A Sketch Book*. Swan Sonnenschein, London, 1887.

Lockhart, J. G. & Woodhouse, C. M. *Rhodes*. Hodder & Stoughton, London, 1963.

McDonald, J. G. *Rhodes: A Life*. Chatto & Windus, London, 1941.

Macmillan, Mona. *Sir Henry Berkly*. Balkema, Cape Town, 1970.

Mathews, J. W. *Incwadi Yami*. Sampson Low, London, 1887.

Maurois, André. *Cecil Rhodes*. Collins, London, 1953.

Michell, Sir Lewis. *The Life of the Right Hon. Cecil John Rhodes*. Edward Arnold, London, 1910.

Millin, S. G. *Rhodes*. Chatto & Windus, London, 1933.

Murray, R. W. *South African Reminiscences*. Juta, Cape Town, 1894.

Payton, Charles. *The Diamond Diggings of South Africa*. Horace Cox, London, 1872.

Phillips, Mrs L. *Some South African Recollections*. Longmans, London, 1899.

Phillips, Lionel. *Some Reminiscences*. Hutchinson, London, 1924.

Plomer, W. *Cecil Rhodes*. Nelson, London, 1933.

Pringle, Thomas. *Narrative of a Residence in South Africa* (Reprint). Struik, Cape Town, 1966.

Raymond, E. T. *Portraits of the Nineties*. T. Fisher Unwin, London, 1921.

Raymond, Harry. *B. I. Barnato: A Memoir*. Isbister & Co., London, 1897.

Rosenthal, E. *Other Men's Millions*. Timmins, Cape Town (n.d.).

— *Gold! Gold! Gold!* Collier Macmillan, Johannesburg, 1970.

Sauer, Hans. *Ex-Africa*. Geoffrey Bless, London, 1937.

Stead, W. T. *The Last Will and Testament of Cecil John Rhodes*. London, 1902.

Taylor, J. B. *A Pioneer Looks Back*. Hutchinson, London, 1939.

Trollope, Anthony. *South Africa* (Abridged edition). Longmans, London, 1938.

Warren, Sir Charles. *On the Veldt in the Seventies*. Isbister & Co., London, 1902.

Weinthal, Leo. *Memories, Mines and Millions*. Simpkin Marshall Ltd., London, 1929.

Williams, A. E. *Some Dreams Come True*. Timmins, Cape Town, 1948.

Williams, Basil. *Cecil Rhodes*. Constable, London, 1921.

Williams, G. F. *The Diamond Mines of South Africa*. Macmillan, London, 1902.

Wilmot, A. *The Life and Times of Sir Richard Southey*. Sampson Low, London, 1904.

Wilson, G. H. *Gone Down the Years*. Timmins, Cape Town, 1947.

Wilson, Lady Sarah. *South African Memories*. Edward Arnold, London, 1909.

Wirgman, A. T. *Storm and Sunshine in South Africa*. Longmans, London, 1902.

'Vindex'. *Cecil Rhodes: His Political Life and Speeches*. G. Bell, London, 1900.

OTHER BOOKS CONSULTED

Cartwright, A. P. *Valley of Gold*. Timmins, Cape Town, 1962.

— *Gold Paved the Way*. Macmillan, London, 1967.
— *The First South African*. Purnell, Cape Town, 1971.
Chilvers, H. A. *Out of the Crucible*. Juta, Johannesburg, 1948.
— *The Seven Lost Trails of Africa*. Dassie Books, Johannesburg, 1949.
Collier, Joy. *The Purple and the Gold*. Longmans, London, 1965.
Couper, J. R. *Mixed Humanity*. W. H. Allen, London (n.d.).
Crisp, Robert. *The Uitlanders*. Peter Davies, London, 1964.
Currey, R. *Rhodes: A Biographical Footnote*. Carmelite Press, Cape Town
    (n.d.).
Fitzpatrick, J. P. *The Transvaal From Within*. Heinemann, London, 1900.
Fort, G. S. *Chance or Design? A Pioneer Looks Back*. Robert Hale, London, 1942.
Hahn, Emily. *Diamond*. Weidenfeld & Nicholson, London, 1956.
Hammond, John Hays. *Autobiography*. Farrar & Rinehart, New York, 1935.
Harris, Sir David. *Pioneer, Soldier and Politician*. Sampson Low, London, 1931.
Herrman, Louis. *A History of the Jews in South Africa*. South African Jewish
    Board of Deputies, Cape Town, 1935.
Hutchinson, G. T. *Frank Rhodes: A Memoir*. Privately Printed, 1908.
Jacobsson, D. *Fifty Golden Years of the Rand*. Faber & Faber, London, 1936.
Jeppe, Carl. *Kaleidoscopic Transvaal*. Chapman & Hall, London, 1906.
Lippert, Marie. *The Travel Letters of Marie Lippert 1891* (translated by Eric
    Rosenthal). Friends of the South African Library, Cape Town, 1960.
MacNeill, J. G. Swift. *What I Have Seen and Heard*. Arrowsmith, London, 1925.
McNish, J. T. *The Road to El Dorado*. Struik, Cape Town, 1968.
— *Graves and Guineas*. Struik, Cape Town, 1969.
— *The Glittering Road*. Struik, Cape Town, 1970.
Pakenham, Elizabeth. *Jameson's Raid*. Weidenfeld & Nicholson, London, 1961.
Poel, Jean van der. *The Jameson Raid*. Oxford University Press, 1951.
Ronan, Barry. *Forty South African Years*. Heath Cranton, London, 1923.
Rouillard, Nancy. *Matabele Thompson*. Faber & Faber, London, 1936.
Scully, W. C. *Reminiscences of a South African Pioneer*. T. Fisher Unwin,
    London, 1913.
— *Further Reminiscences of a South African Pioneer*. T. Fisher Unwin, London,
    1913.
Statham, F. Reginald. *South Africa As It Is*. T. Fisher Unwin, London, 1897.
Taylor, W. P. *African Treasures: Sixty years among diamonds and gold*. John
    Long, London, 1932.
Vulliamy, C. E. *Outlanders*. Cape, London, 1938.
Walker, E. A. *History of South Africa*. Longmans, London, 1957.

For Newspapers and Magazines
See Reference Notes

INDEX